Salesforce Process Builder Quick Start Guide

Build complex workflows by clicking, not coding

Rakesh Gupta

BIRMINGHAM - MUMBAI

Salesforce Process Builder Quick Start Guide

Commissioning Editor: Richa Tripathi
Acquisition Editor: Noyonika Das
Content Development Editor: Mohammed Yusuf Imaratwale
Technical Editor: Shweta Jadhav
Copy Editor: Safis Editing
Project Coordinator: Hardik Bhinde
Proofreader: Safis Editing
Indexer: Rekha Nair
Graphics: Jason Monteiro
Production Coordinator: Arvindkumar Gupta

First published: July 2018

Production reference: 1180718

Published by Packt Publishing Ltd.
Livery Place
35 Livery Street
Birmingham
B3 2PB, UK.

ISBN 978-1-78934-431-8

www.packtpub.com

This book is dedicated to my parents Kedar Nath Gupta and Madhuri Gupta for their sacrifices and for exemplifying the power of determination

– Rakesh Gupta

Mapt

Foreword

If you have ever had a question about Salesforce Process Automation and searched the web or the Salesforce Trailblazer Community for answers, you have likely already come across Rakesh Gupta. Rakesh runs the automationchampion.com blog, co-hosts the Salesforce Automation Hour webinar series, is a four-time Salesforce MVP, leads the Mumbai Salesforce User Group and contributed to more than 100 "best-answers' in the Trailblazer Community. In short, he is the definition of an expert on Salesforce Process Automation and I would be hard-pressed to find someone better suited to author this book.

At Salesforce we know that customers expect seamless experiences, and customer expectations are rising rapidly now that we are in the fourth industrial revolution. However, providing seamless, automated customer experiences has historically been challenging, time-consuming and code heavy. This is why Salesforce has invested in point-and-click tools like Process Builder and Flow to enable you to declaratively configure process logic, integrate multiple systems and deliver rich user experiences. This book is an invaluable resource to learn Salesforce Process Builder so you too can deliver seamless, automated experiences to your customers.

In my own experience, as I have worked with customers, admins and MVPs on Process Builder and Flow, I have found the content that Rakesh has authored to be a great resource – a tailor-made set of answers to many common questions and best practices about these Salesforce platform services. His dedication to sharing his knowledge and inspiring the Salesforce Ohana is incredible and it's an immense pleasure for me, as the product manager for Process Automation at Salesforce, to write the foreword for this book.

Arnab Bose
VP, Product Management, Salesforce.

Contributors

About the author

Rakesh Gupta is best known as an automation champion in the Salesforce ecosystem. He has written over 150 articles on Visual Workflow and Process Builder to show how someone can use Visual Workflow and Process Builder to minimize code usage. He is one of the Visual Workflow and Process Builder experts in the industry. He has trained more than 700 individual professionals around the globe and conducted corporate training. Currently, Rakesh is working as a Salesforce solution architect consultant.

First and foremost, I would like to thank my parents, Kedar Nath Gupta and Madhuri Gupta, and my sister, Sarika Gupta, for being patient with me for taking yet another challenge. I would also like to thank my friend, Meenakshi Kalra, for helping me while I was writing this book. Special thanks to Arnab Bose for writing the foreword.

Packt is searching for authors like you

If you're interested in becoming an author for Packt, please visit `authors.packtpub.com` and apply today. We have worked with thousands of developers and tech professionals, just like you, to help them share their insight with the global tech community. You can make a general application, apply for a specific hot topic that we are recruiting an author for, or submit your own idea.

Table of Contents

Preface

Salesforce Management System is an information system used in Customer Relationship Management (CRM) to automate business processes like sales and marketing. Force.com built a powerful application, called Process Builder, to automate business processes.

Salesforce Process Builder Quick Start is a practical guide to turbo-charge your ability to learn process builder; once mastered, you will be able to develop custom applications in Salesforce with minimal code usage.

The book starts with an introduction to Process Builder. It focuses on the building blocks of creating Processes. Then, it will discuss varied applications of Process Builder to develop streamlined solutions. You will learn how to easily automate business processes and tackle complex business scenarios using Processes. The book explains the workings of the Process Builder so that you can create reusable processes. It also covers how you can migrate the existing Workflow Rules to one Process Builder.

By the end of the book, you will get a clear understanding of how to use Flows and Process Builder to optimize code usage.

Who this book is for

This book is intended for those who want to use Process Builder to automate their business requirements by click, not code. Whether you are new to Salesforce or you are a seasoned expert, you want to master both Flow and Process Builder. Since Salesforce maintains an incredibly user-friendly interface, no previous experience in computer coding or programming is required. The things that you do require are your brain, your computer with a modern web browser, a free Salesforce developer org, and just basic knowledge of Salesforce.

What this book covers

`Chapter 1`, *Getting Started with Lightning Process Builder*, starts with basic knowledge of Salesforce Process Builder. We will then pick a few business examples and see how to use Process Builder instead of Apex code to solve them. We'll then discuss the benefits of using Lightning Process Builder. You will also get an overview of the Process canvas and its features.

`Chapter 2`, *Deploying, Distributing, and Debugging Processes*, serves as the climax of the book, where you will learn how to debug your process. We will also cover how to distribute your Process and setup recipients to receive process error emails.

`Chapter 3`, *Building Efficient and Performance-Optimized Processes*, helps you understand Process Builder and its concepts, such as how to use Custom Metadata Types, Custom Permissions, and Custom Labels with Process Builder. We will also cover some key concepts, such as using multiple groups of actions and how to bypass Process Builder for a user or profile.

To get the most out of this book

All you need to get the most out of this book is your brain, your computer with a modern web browser, and a free Salesforce developer org. You can sign up for a free developer org at `https://developer.salesforce.com/signup`.

Download the color images

We also provide a PDF file that has color images of the screenshots/diagrams used in this book. You can download it here: `https://www.packtpub.com/sites/default/files/downloads/SalesforceProcessBuilderQuickStartGuide_ColorImages.pdf`.

Conventions used

There are a number of text conventions used throughout this book.

`CodeInText`: Indicates code words in text, database table names, folder names, filenames, file extensions, pathnames, dummy URLs, user input, and Twitter handles. Here is an example: "Enter `Add record to Chatter group` as the action name, in this case."

Bold: Indicates a new term, an important word, or words that you see onscreen. For example, words in menus or dialog boxes appear in the text like this. Here is an example: "Select **System info** from the **Administration** panel."

Warnings or important notes appear like this.

Tips and tricks appear like this.

Get in touch

Feedback from our readers is always welcome.

General feedback: Email `feedback@packtpub.com` and mention the book title in the subject of your message. If you have questions about any aspect of this book, please email us at `questions@packtpub.com`.

Errata: Although we have taken every care to ensure the accuracy of our content, mistakes do happen. If you have found a mistake in this book, we would be grateful if you would report this to us. Please visit `www.packtpub.com/submit-errata`, selecting your book, clicking on the Errata Submission Form link, and entering the details.

Piracy: If you come across any illegal copies of our works in any form on the Internet, we would be grateful if you would provide us with the location address or website name. Please contact us at `copyright@packtpub.com` with a link to the material.

If you are interested in becoming an author: If there is a topic that you have expertise in and you are interested in either writing or contributing to a book, please visit `authors.packtpub.com`.

Reviews

Please leave a review. Once you have read and used this book, why not leave a review on the site that you purchased it from? Potential readers can then see and use your unbiased opinion to make purchase decisions, we at Packt can understand what you think about our products, and our authors can see your feedback on their book. Thank you!

For more information about Packt, please visit `packtpub.com`.

1

Getting Started with Lightning Process Builder

This chapter will start with an overview of Process Builder and its benefits, followed by an illustration of how to use Process Builder to automate various business processes. By the end of the chapter, you will have learned how to use Lightning Process Builder's tools and many of its different elements.

In the next chapter, you will be briefed on various tips and tricks related to Process Builder, including how to debug and distribute a Process. You will also see different ways to streamline a sales process and to automate a business process. In the last chapter, we will go over how to create reusable processes using Process Builder, how to bypass Process Builder for a set of users, and much more.

The following topics will be covered in this chapter:

- An overview of Process Builder
- The differences between Process Builder and other tools
- Creating applications with Process Builder
- Various use cases of Process Builder
- Limitations of Process Builder

Just to remind you, we will use Lightning Experience for all chapters.

An overview of Process Builder

Lightning Process Builder and Process Builder are the same tool. Process Builder provides a way to automate business processes. In other words, it is the upgraded version of the Workflow Rule. Whenever you create a process, the system automatically creates a **Flow,** and a **Flow Trigger** to call the Flow. This happens behind the scenes, and the user doesn't need to interact with the shadow Flows. The Workflow Rule has several limitations. It doesn't allow you to update child records. Also, it doesn't allow you to post to Chatter, create a child record on a specific action, or automatically submit a record for approval.

To overcome these limitations, Salesforce introduced Process Builder in its Spring 2015 release. There are a few advantages to Process Builder, which are as follows:

- It allows you to create a complete process on a single screen, unlike in Workflow Rules, where you have to move from screen to screen to create a complete rule.
- Its visual layouts allow you to create a complete process using point-and-click.
- It helps you to minimize Apex code usage.
- It allows you to call Apex from Process Builder, where Apex is still required.
- It also allows you to create multiple scheduled actions for the criteria from Process Builder.
- You can easily reorder process criteria with drag and drop.
- It is also possible to execute multiple criteria of a process.

As of the Summer 2018 release, Process Builder runs in **system mode,** so object- and field-level permissions will be ignored for the user that triggers the process. Visual Workflow runs in **user mode**, which means that at runtime, the user that triggers the Flow, their access on the object, and field the level will be counted. However, if a process is launching a Flow, the whole automation will run in system mode. Let's look at an example; suppose that you are trying to update the opportunity **Next Step** field:

1. **If you are using Process Builder**: If the running user doesn't have access to the **Next Step** field, Process Builder will be able to update it.
2. **If you are using Flow (a custom button to call a Flow)**: If the running user doesn't have access to the **Next Step** field, they will get an error.
3. **If you are using Flow to achieve the same thing, and you are using Process Builder to auto-launch the Flow**: If the running user doesn't have access to the **Next Step** field, then the Flow will be able to update it.

If any of the actions fail at runtime, the entire transaction will fail, and an error message will be displayed. There are some exceptions to this and settings to work around it, which we will discuss in `Chapter 2`, *Deploying, Distributing and Debugging your Process*.

Business problems

As a Salesforce administrator or developer, you may receive multiple requirements from a business, to streamline sales or support processes. If something can't be achieved using Workflow Rule, then you may have to use Apex code to automate it. Let's look at a business scenario.

The use case is as follows: Sara Bareilles is working as a Salesforce administrator at Universal Containers. She has a requirement to auto-update the related contacts **Other Phone** field with the account Phone, once the account has been activated.

There are several ways to fulfill the preceding business requirement:

1. To fulfill this business requirement, we could create a Flow and embed it in a Visualforce page. Then, we could use it as an inline Visualforce page in an account page layout.
2. Since we can't achieve the business requirement using a Workflow Rule, the next possibility is to use an Apex trigger. A developer writes an Apex trigger on an **Account** object, to update all contacts when the account is activated.
3. You can also use Process Builder. We will discuss that in detail later in this chapter.

Browser requirements for Process Builder

Process Builder is available for Lightning Essentials and Professional (with a limited number of processes), Lightning Enterprise, Lightning Unlimited, and Lightning Developer editions. You can access Process Builder on any platform. The requirements are as follows:

- The most recent version of Google Chrome
- The most recent version of Mozilla Firefox
- The most recent version of Safari
- The most recent version of Internet Explorer

Process Builder is 508-compliant, which means that all users, regardless of disability status, can access Process Builder, with one exception:

```
They can close modal dialogs using the ESC key on their keyboard, but they
can't close side panels by using the ESC key.
```

An overview of the Process Builder user interface

Process Builder is a tool that allows you to implement business requirements by creating processes (without any code). It has almost all of the features that are offered by Workflow Rule, and it also contains some new features, such as **Post to Chatter**, **Launch a Flow**, **Create a Record**, **Update Records**, and **Submit for Approval**. From now on, we will use Lightning Experience to create or manage processes using Process Builder, and to create or manage Visual Workflows. The Process Builder user interface has different functional parts, which are shown in the following screenshot:

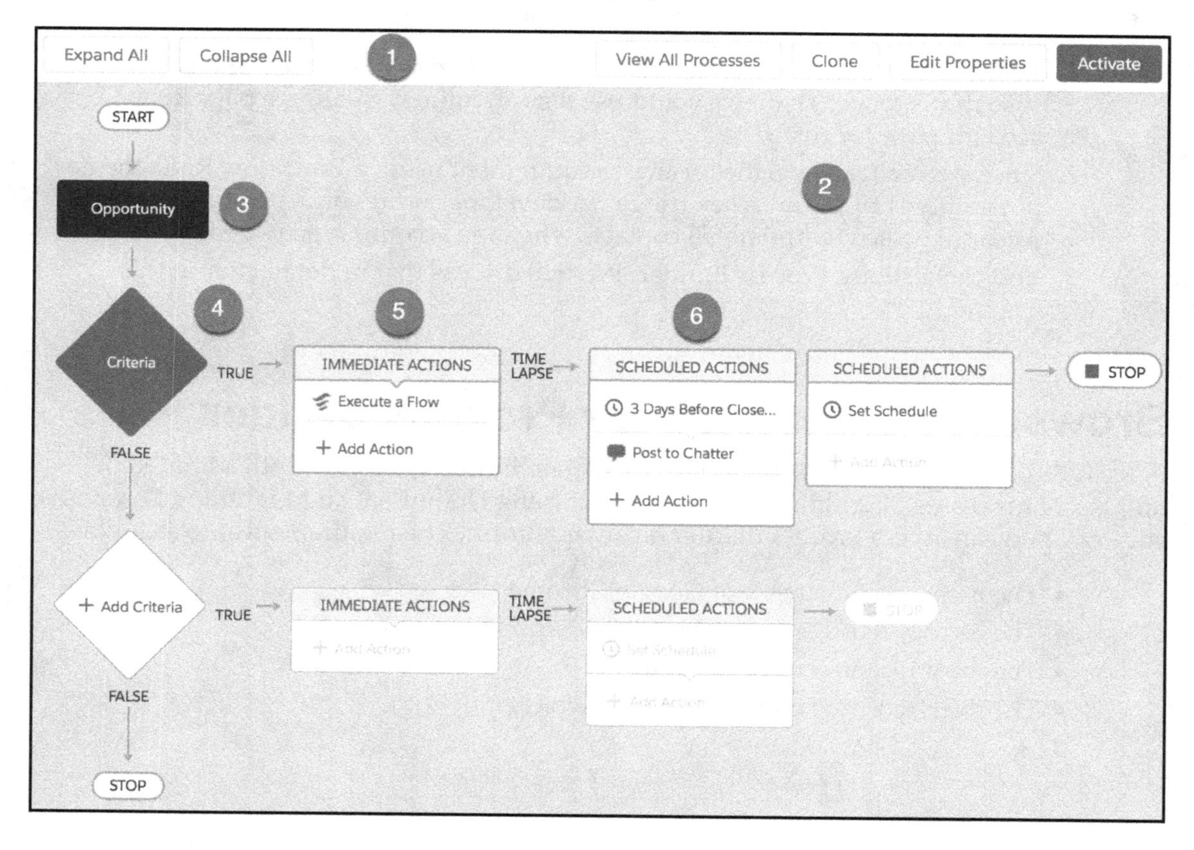

The different functional parts of the user interface of Process Builder are as follows:

1. **Button bar**: The following are the buttons available in the button bar:
    - **Activate**: Use this button to activate your process. You can't make any changes once a process is activated.
    - **Deactivate**: This button is available on the button bar only if the process is activated. Use this button to deactivate a process.
    - **Edit Properties**: This will show you the **Process Name**, **API Name**, and **Description** fields of your process. It allows you to change the process name and description, as long as the process is not activated. You can't change the **API Name** field after you've saved it for the first time. The **Properties** window will look as follows:

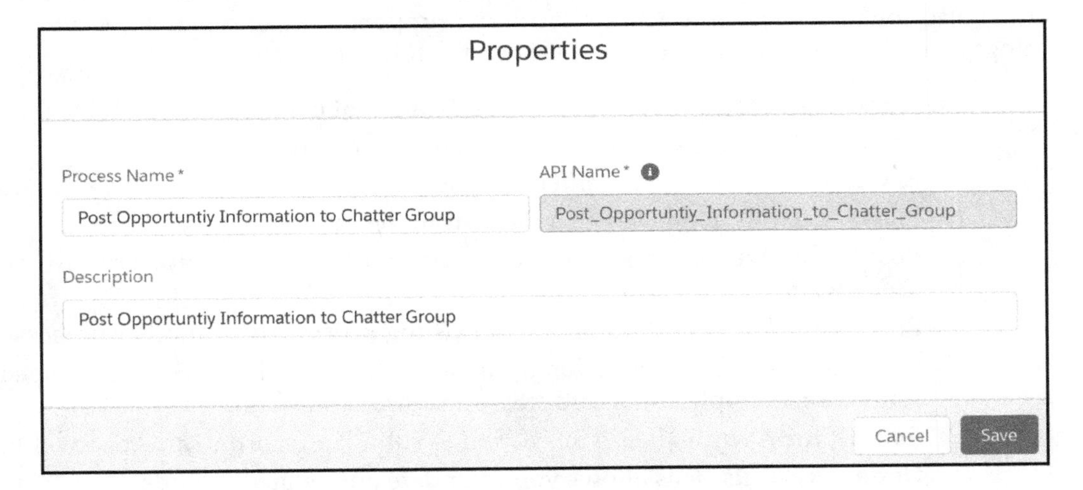

    - **Clone**: The **Clone** button lets you make a copy of the current process. You will be given two **Clone As** options – **Version of current process** and **A new process**.
    - **View All Processes**: When you click on this button, it will redirect you to the **Processes Management** page. From there, you can see all of the processes created in the current Salesforce organization.
    - **Collapse All**: Collapse all actions on the canvas.
    - **Expand All**: Expand all actions on the canvas.

2. **Process canvas**: This is the main area, where you can use point-and-click to develop a process. To edit any element on the process canvas, double-click on it.
3. **Add object**: Select the object upon which you want to create a process, and choose the evaluation criteria (the changes that will cause the process to run).
4. **Add criteria**: Use this to define the criteria and set the filter conditions.
5. **IMMEDIATE ACTIONS**: Use this to define immediate actions for the process.
6. **SCHEDULED ACTIONS**: Use this to define scheduled actions for the process.

Actions available in Process Builder

Process Builder can perform almost all of the actions that are available for Workflow Rules, and it also contains some new actions. It doesn't support outbound messages, among other things. With Process Builder, you can perform the following actions:

- **Apex**: This allows you to call an Apex class that contains an invocable method.
- **Create a Record**: Using this, you can create a record.
- **Email Alerts**: Use this to send email alerts.
- **Flows**: Use this action to call a Flow from the Process.
- **Post to Chatter**: Use this to post a textpost on a Chatter group, a record, or a user's wall.
- **Processes**: Use this action to call an existing process from another process.
- **Quick Actions**: Use this action to call a Chatter global action or object specific action; for example, log a call, create a record, and so on.
- **Submit for Approval**: Use this action to submit a record to an Approval Process.
- **Update Records**: This allows you to update any related records.

In this and the following chapters, we will see each action in detail.

Differences between Process Builder and other tools

Salesforce offers various tools to automate business processes; for example, Visual Workflow, Workflow Rule, and Process Builder. So, it is necessary to understand the differences between these tools, and when to use which one. The following table describes the differences between these tools:

	Workflow	**Flow**	**Process Builder**
Visual designer	Not available	Available	Available
Starts when	A record is created or edited	• The user clicks on a custom button/link • A process starts • Apex is called • Inline Visualforce page • The user accesses a custom tab	A record is created or edited
Supports time-based actions?	Yes	Yes	Yes
Call Apex code?	No	Yes	Yes
Create records	Only task	Yes	Yes
Invoke processes	No	No	Yes
Update records?	Yes, but only fields from the same record or parent (in case of master-detail relationship)	Yes, any record	Yes, any related record
Delete records?	No	Yes	No
Launch a Flow?	No	Yes	Yes
Post to Chatter?	No	Yes	Yes, but only textpost
Send an email?	Yes	Yes	Yes
Submit for approval?	No	Yes	Yes
Send outbound messages?	Yes	No	No
Supports user interaction?	No	Yes	No
Version control?	No	Yes	Yes

	Workflow	**Flow**	**Process Builder**
Supports user input at runtime?	No	Yes (through screen elements)	No
Supports unauthenticated access?	No	Yes (through `Force.com` sites)	No
Can pause on runtime?	No	Yes	No
Allows modification	After deactivation, you can modify the Workflow Rule	Once Flow is activated, you can't modify it, and the same applies after deactivation	Once process is activated, you can't modify it, and the same applies after deactivation
Delete	Once Workflow is deactivated, you can immediately delete it	Once Flow is deactivated, you can immediately delete it	Once process is deactivated, you can immediately delete it
Version Control	No	Yes (number of versions you can create for a Flow is 50)	Yes (number of versions you can create for a process is 50)

Creating applications with Process Builder

Before Process Builder was available, knowledge of Apex and Visualforce was required to automate complex business processes in Salesforce. After completing this chapter, you will have a clear idea of how to automate business processes using Process Builder, and will minimize your need for Apex code. Now, we will discuss how to use Process Builder to automate business processes. From now on, we will use Lightning Experience to create or manage processes using Process Builder, and to create or manage Visual Workflows.

Hands on 1 – auto-create a child record

Businesses commonly require the auto-creation of a child record whenever a parent record gets created. For example, as soon as an account gets activated, we need to auto-create an opportunity for that account, and we need to set the opportunity close date to the last date of the current quarter. To satisfy this type of business requirement, a developer normally writes an Apex trigger, but there are a few other ways to achieve it, without writing code:

- Using Process Builder
- Using a combination of Flow and Process Builder
- Using a combination of Flow and an Inline Visualforce page, on the account detail page

We will use Process Builder to solve this business requirement.

Let's consider a business scenario. Suppose that Joe Thompson is working as a system administrator at Universal Containers. He has received a requirement from the management to auto-create a contract as soon as an account gets created in Salesforce, and to auto-populate these values in the new contract:

- **Contract term (months)**: 12
- **Contract start date**: Account created date, + 90 days
- **Status**: Draft
- **Auto-relate it with a new account**

Creating a process

To solve the preceding business requirement, we will use Process Builder. In the runtime process, we will auto-create a contract record whenever a new account record gets created. To do this, follow these instructions:

1. In Lightning Experience, click on **Setup** (gear icon) | **Setup** | **PLATFORM TOOLS** | **Process Automation** | **Process Builder**; click on the **New** button, and enter the following details:
    - **Process Name**: Enter the name of the process. Enter `Auto create new Contract` as the **Process Name**. It must be within 255 characters.
    - **API Name**: This will be auto-populated, based on the name. It must be within 77 characters.

- **Description**: Write some meaningful text, so that other developers or administrators can easily understand why this process was created.
- **This process starts when**: This allows you to select when you want to start your process. The following options are available:
    - **A record changes**: Select this option if you want to start your process when a record is created or edited.
    - **It's invoked by another process**: Select this option if you want to invoke your process from another process. This allows you to create an **invocable process**. An invocable process is a process that starts when another process invokes it. In `Chapter 3`, *Building Efficient and Performance-Optimized Processes,* we will discuss a few use cases in detail.

In this case, select **A record changes**. The fields should appear as follows:

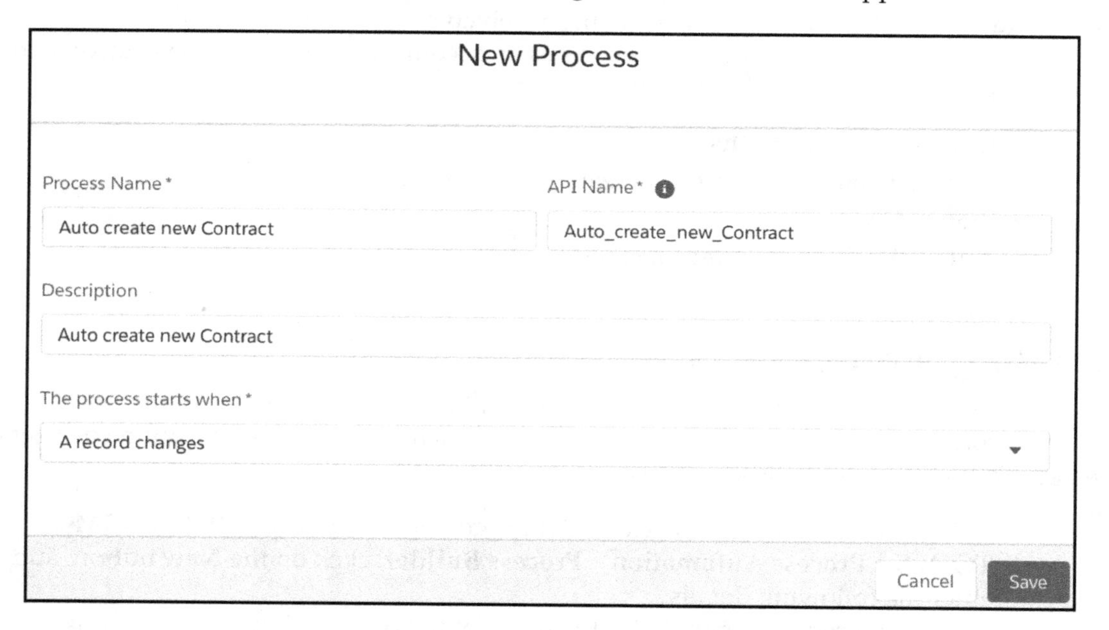

2. Click on the **Save** button when you have finished. It will redirect you to the process canvas, which will allow you to create the process by clicking, not code.

Adding an object and evaluation criteria

Once you are done defining the process properties, the next task is to select the object upon which you want to create a process and define the **evaluation criteria**:

1. Click on the **Add Object** node, as shown in the following screenshot:

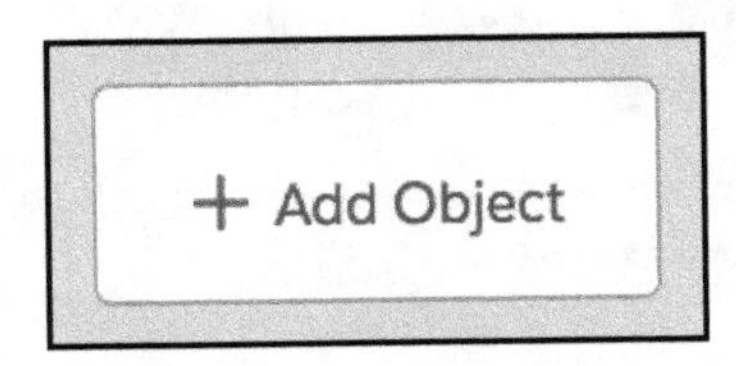

2. This will open a window on the right-hand side, where you will have to enter the following details:

    - **Object**: Start typing, and then select the **Account** object.
    - **Start the process**: For **Start the process**, select **only when a record is created**. This means the process will fire only at the time of record creation.
    - **Recursion – Allow process to evaluate a record multiple times in a single transaction?**: Select this checkbox only when you want the process to evaluate the same record up to five times in a single transaction. It might reexamine the record, because a process, Workflow Rule, or Flow may have updated the record in the same transaction. In this case, leave the box unchecked.

- The window should appear as shown in the following screenshot:

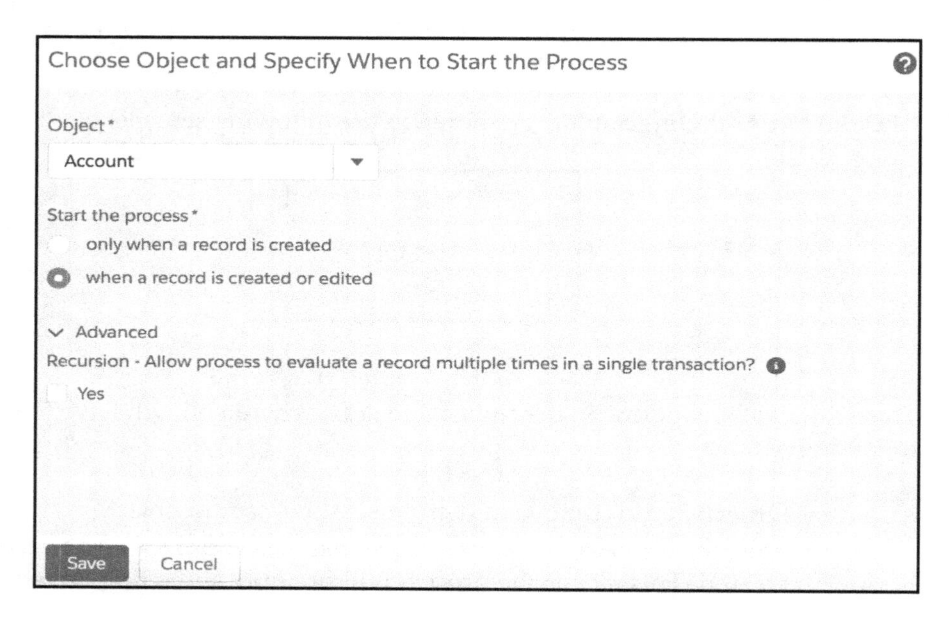

3. Once you are done, click on the **Save** button. Just as in Workflow Rule, once you have saved the panel, it doesn't allow you to change the selected object.

Adding process criteria

After defining the evaluation criteria, the next step is to define the process criteria. This is similar to the rule criteria in Workflow Rule. Once the process criteria are true, the process will execute with the associated actions:

1. To define the process criteria, click on the **Add Criteria** node, as shown in the following screenshot:

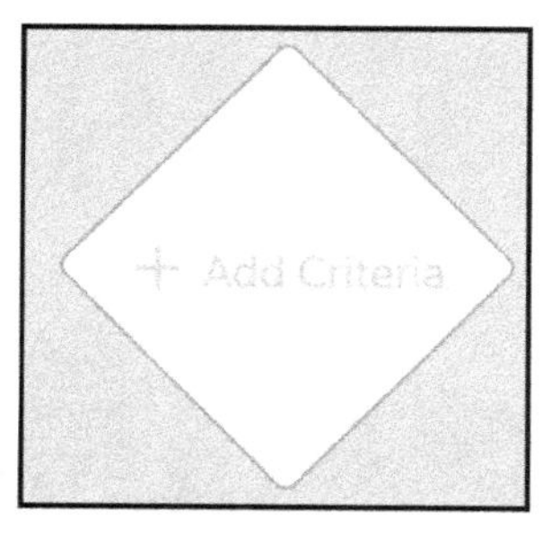

2. Now, enter the following details:
    - **Criteria Name**: Enter a name for the criteria node. Enter `Always` in **Criteria Name**.
    - **Criteria for Executing Actions**: Select the type of criteria that you want to define. You can select **Formula evaluates to true,** or **Conditions are met** (a filter to define the process criteria), or **No criteria-just execute the actions!** In this case, select **No criteria-just execute the actions!** This means that the process will fire for every condition. The window should look as follows:

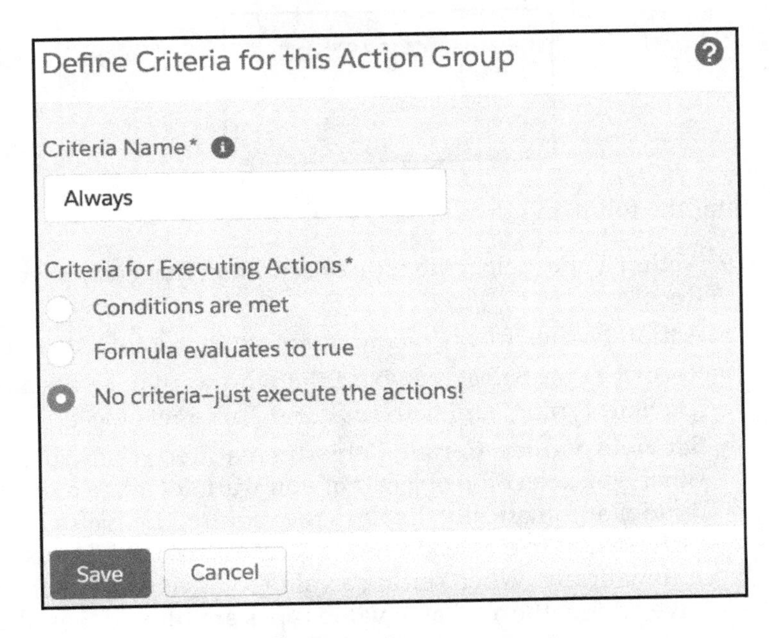

3. Once you are done, click on the **Save** button.

Adding an action to a process

Once you are done with the process criteria node, the next step is to add an immediate action to create a contract. For this, we will use the **Create a Record** action, available in Process Builder. You can add multiple immediate and scheduled actions for a particular criteria node. The only limitation is the maximum number of criteria nodes that you can add in one process, which is `200`. Immediate actions are executed as soon as the evaluation criteria are met.

Scheduled actions are executed at a specified time. We will add an immediate action with the following steps:

1. We need to add one immediate action to auto-create a record. For this, we will use the **Create a Record** action, available in Process Builder. Click on **Add Action,** available under **IMMEDIATE ACTIONS**, as shown in the following screenshot:

2. Now, enter the following details:

    - **Action Type**: Select the type of action. In this case, select **Create a Record**.
    - **Action Name**: Enter `Create Contract record` in **Action Name**.
    - **Record Type**: Select the object that you want to create a record for. Start typing, and then select the **Contract** object.
    - **Set Field Values**: Certain fields are required when you create a record. When you select the object that you want to create a record for, Process Builder automatically displays the required fields for that record in each row. In this case, a row for **Account ID** and **Status** shows automatically. When setting a value for a given field or a field you have added, the available value types are filtered, based on the field that you have selected. The available value types are:
        - **Currency**: You can manually enter a currency value.
        - **Boolean**: This allows you to choose a true or false Boolean value.
        - **DateTime or Date**: You can manually enter a **DateTime** or **Date** value.
        - **Formula**: You can easily create a formula, using functions and fields.

- **Global Constant**: This allows you to choose a global constant to set a value to null or an empty string. For example, choose `$GlobalConstant.Null` or `$GlobalConstant.EmptyString`
- **ID**: You can manually enter a Salesforce ID.
- **MultiPicklist**: This allows you to choose one or more multi-select picklist values.
- **Number**: You can manually enter a number value.
- **Picklist**: This allows you to choose a picklist value.
- **Reference**: This allows you to choose a value, based on a field on the record or on a related record.
- **String**: You can manually enter a string value.

To select the fields, you can use **field picker**. To enter the values, use the **text entry** field and map the fields according to the following screenshot:

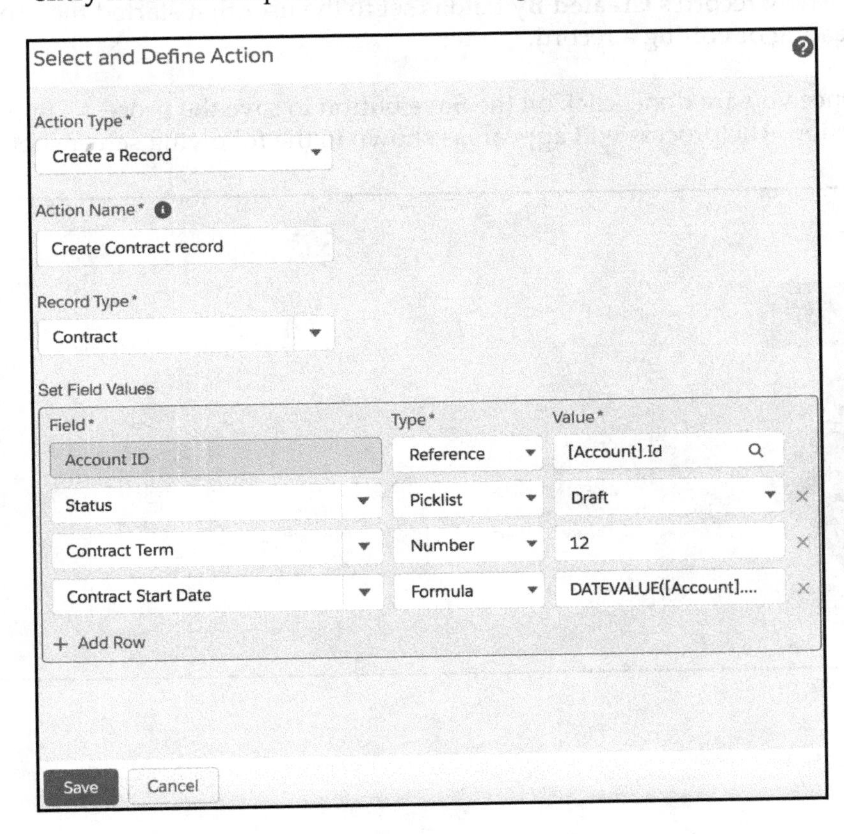

Use the formula to set the **Contract Start Date** dynamically, based on the preceding requirement; in this case, you should use the account created date + 90 days. Use **Function** to select **DATEVALUE** (it returns a date value for a **DateTime**), use the **Field** picker to select the **Created Date** field, use **Operator** to select **+**, and then enter `90`, as shown in the following screenshot:

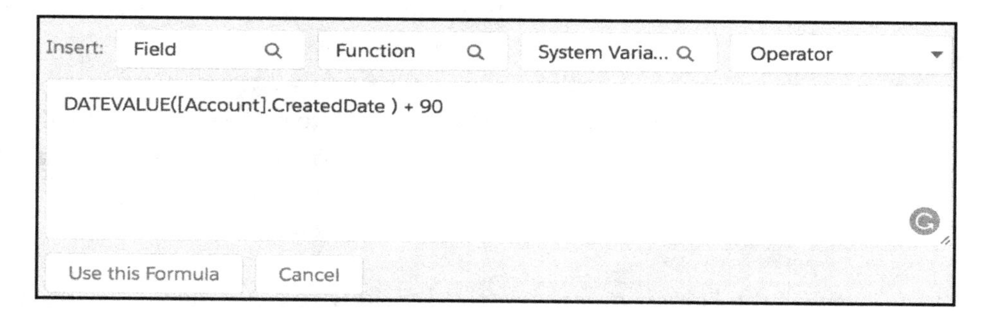

The new record's **Created By** field is set to the user that started the process by creating or editing a record.

3. Once you are done, click on the **Save** button to save the process's immediate action. The process will appear, as shown in the following screenshot:

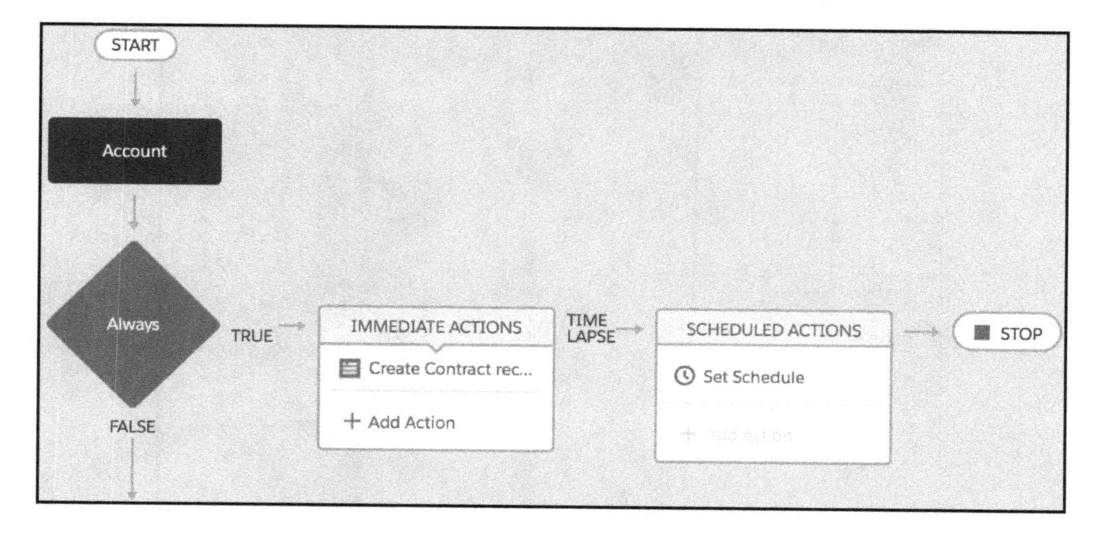

Activating a process

As soon as you are done with the process creation, the next step is to activate it:

1. To activate a process, click on the **Activate** button, available on the button bar, as shown in the following screenshot:

2. A warning message will appear on the screen. Read it carefully, and then click on the **Confirm** button, as shown in the following screenshot:

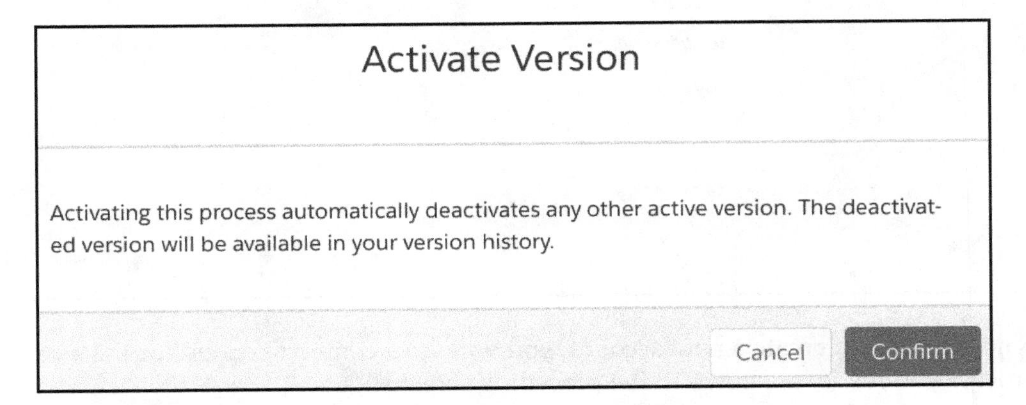

3. When you click on the **Confirm** button, it will activate the process.

For activation, a process must have an action added to it. If a process doesn't have an action added to it, the activate button won't be clickable. After the activation of a process, it's impossible to make any changes to it. If you want to do that, you will have to clone the process, and save it as either a **New Version** or a **New Process**, which we will discuss later:

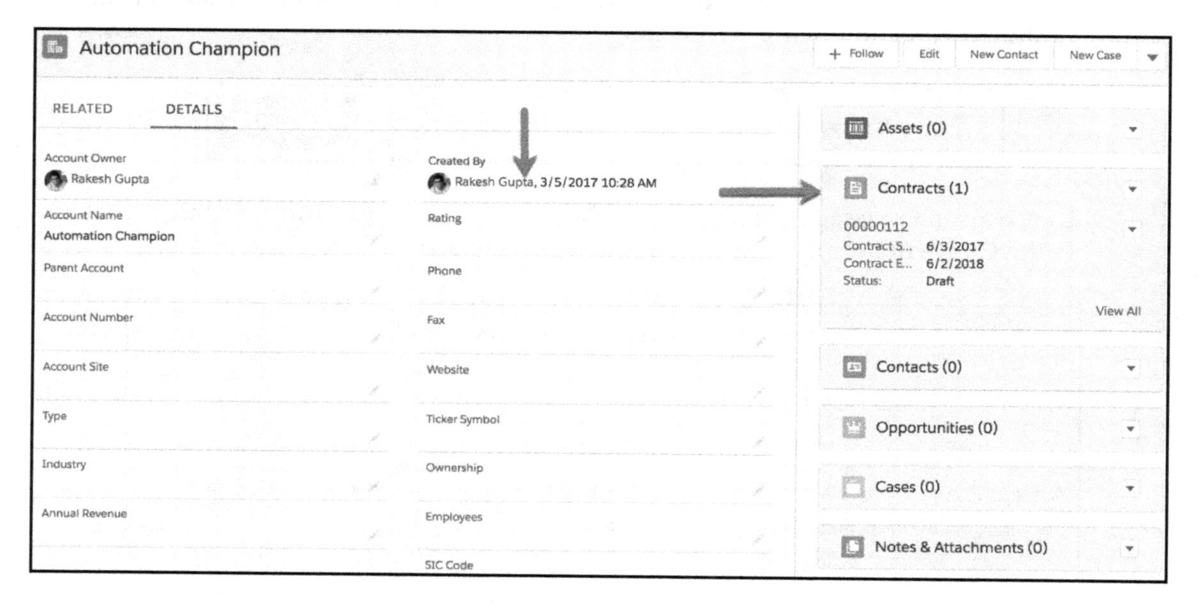

From now on, if you create a new account, you will see a contract created for the account by the process you made, as shown in the preceding screenshot.

Deactivating a process

If some active processes are no longer required by the business, you can deactivate them. To deactivate an activated process, open the process and click on the **Deactivate** button, as shown in the following screenshot:

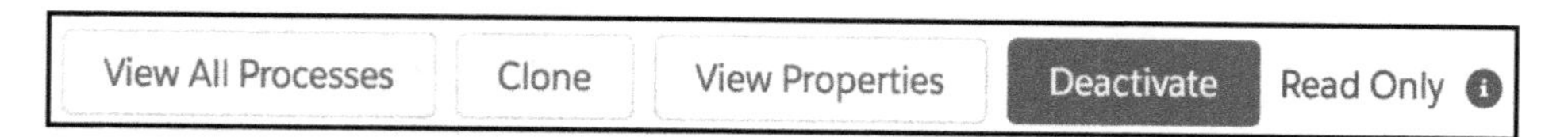

After deactivation, Salesforce stops using the process to evaluate when an account record is created or edited. After you deactivate a process, any scheduled actions will still be there, in a queue for execution.

Deleting a process version

If a process version is no longer in use, you can delete it. You can only delete those process versions that are in the **Draft**, **Inactive**, or **Invalid Draft** status. This means that you can't delete an active process. If you want to delete an active process version, you must deactivate it first, and then delete it. If a process version has scheduled actions, then you can't delete it; in such a case, you must wait until those pending actions have been completed or deleted. To delete a process, follow these instructions:

1. Navigate to **Setup** (gear icon) | **Setup** | **PLATFORM TOOLS** | **Process Automation** | **Process Builder.**
2. This will redirect you to the **Process Management** page. Click on the name of the process whose version you want to delete.
3. Identify the process version that you want to delete, and click on **Delete**, as shown in the following screenshot:

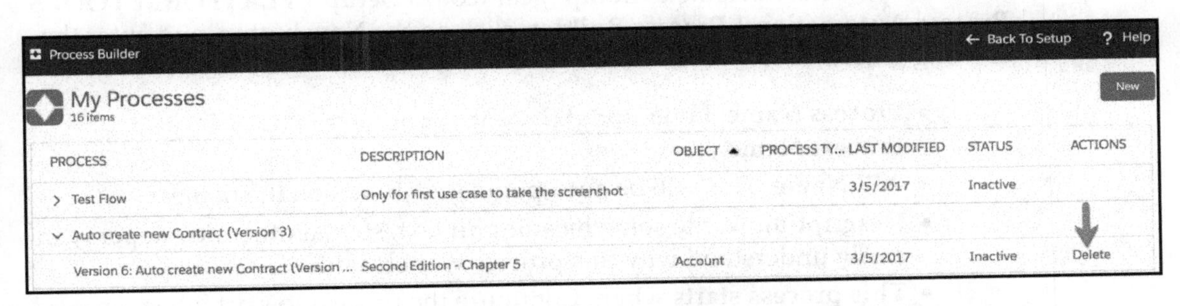

If your process has only one version and you delete that version, the whole process will be removed from the **Process Management** page.

Hands on 2 – auto-update child records

There may be business use cases wherein a customer will want to update child records based on some criteria; for example, auto-updating all related opportunities to **Closed Lost**, if an account is updated to **inactive**. To achieve these types of business requirements, you can use an Apex trigger. You can also achieve these types of requirements using the following methods:

- Process Builder.
- A combination of Flow and Process Builder.
- A combination of Flow and the Inline Visualforce page, on the account detail page.

We will use Process Builder to solve these types of problems.

Let's look at a business scenario. Suppose that Alice Atwood is working as a system administrator at Universal Containers. She has received a requirement that once an account gets activated, the account Phone must be synced with the related contact Asst. Phone field. This means that whenever an account Phone field gets updated, the same phone number should be copied to the related contact's Asst. Phone field.

Follow these instructions to achieve the preceding requirement using Process Builder:

1. First of all, navigate to **Setup** (gear icon) | **Setup** | **PLATFORM TOOLS** | **Objects and Fields** | **Object manager** | **Fields & Relationships,** and make sure that the **Active** picklist is available in your Salesforce organization. If it's not available, create a custom **Picklist** field with the name **Active**, and enter the `Yes` and `No` values.
2. To create a process, navigate to **Setup** (gear icon) | **Setup** | **PLATFORM TOOLS** | **Process Automation** | **Process Builder**, click on the **New** button, and enter the following details:

    - **Process Name**: Enter `Update Contacts Asst Phone` in **Process Name.**
    - **API Name**: This will be auto-populated, based on the name.
    - **Description**: Write some meaningful text, so that other developers can easily understand why this process was created.
    - **This process starts when**: Configure the process to start when a record is created or edited. In this case, select **A record changes.**

The properties window will look as follows:

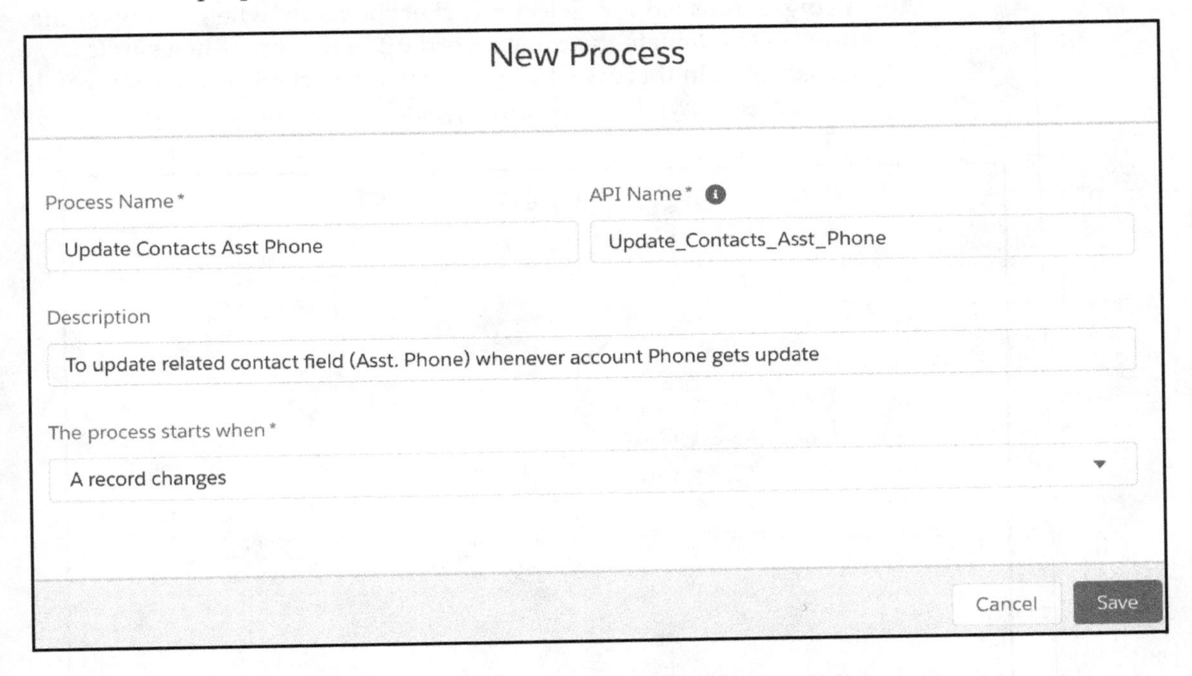

3. Once you are done, click on the **Save** button. It will redirect you to the process canvas, which will allow you to create or modify the process.
4. After defining the process properties, the next task is to select the object upon which you want to create a process and define the evaluation criteria. For this, click on the **Add Object** node. It will open an additional window on the right-hand side of the process canvas screen, where you will have to enter the following details:
    - **Object**: Start typing, and then select the **Account** object.
    - **Start the process**: For **Start the process**, select **when a record is created or edited**. This means that the process will fire every time, irrespective of record creation or update.

- **Recursion - Allow process to evaluate a record multiple times in a single transaction?**: Select this checkbox only when you want the process to evaluate the same record up to five times in a single transaction. In this case, leave the box unchecked. The window will appear as shown in the following screenshot:

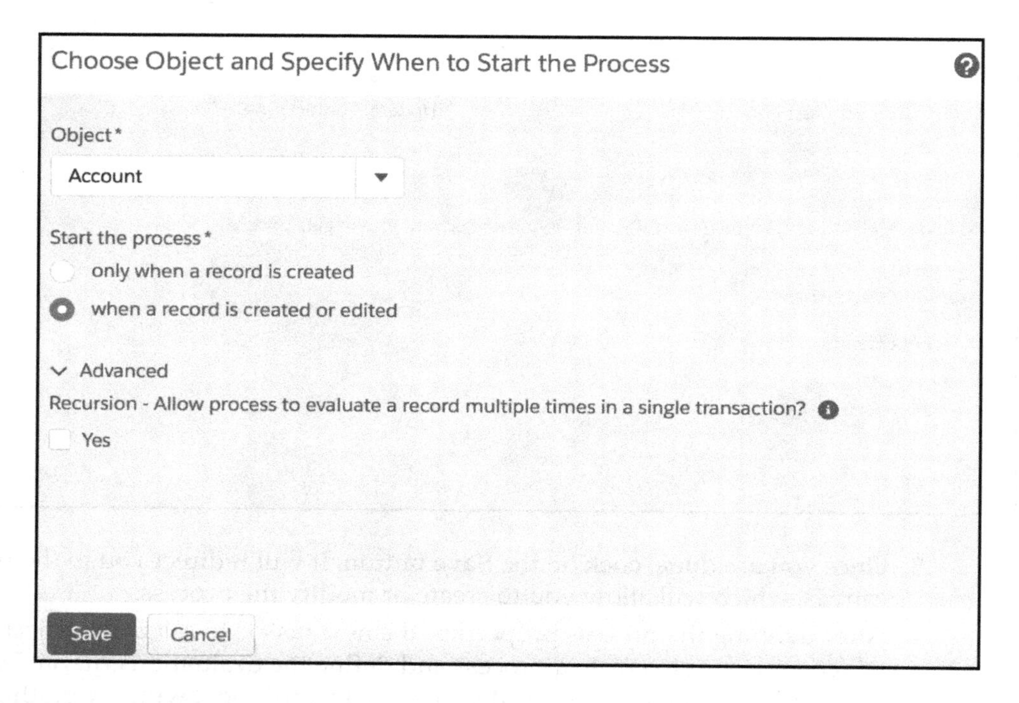

5. Once you are done, click on the **Save** button.
6. After defining the evaluation criteria, the next step is to add the process criteria. Once the process criteria are true, the process will execute the associated actions. To define the process criteria, click on the **Add Criteria** node. It will open an additional window on the right-hand side of the process canvas screen, where you will have to enter the following details:
    - **Criteria Name**: Enter `Update Contacts` in **Criteria Name**.

- **Criteria for Executing Actions**: Select the type of criteria that you want to define. You can select either **Formula evaluates to true,** or **Conditions are met** (a filter to define the process criteria), or **No criteria-just execute the actions!** In this case, select **Conditions are met.**
- **Set Conditions**: This field lets you specify which combination of the filter conditions must be true for the process to execute the associated actions. In this case, select **Active** as **Yes**. This means that the process will fire only when the account is active.
- **Conditions**: In the **Conditions** section, select **All of the conditions are met (AND)**. This field lets you specify which combination of the filter conditions must be true for the process to execute the associated actions.
- Under **Advanced**, select **Yes** to execute the actions only when the specified changes are made. This means that the actions will be executed only if the record meets the criteria now but the values that the record had immediately before it was saved didn't meet criteria. This means that these actions won't be executed when irrelevant changes are made. For example, if the account status is **Active** and a user updates the record by adding a **Shipping Address**, the process won't execute the associated actions. This setting isn't available if:
    - Your process starts only when a record is created.
    - Your process starts when a record is created or edited, and the criteria node doesn't evaluate any criteria.
    - The criteria node evaluates a formula, but the formula doesn't include a reference to the record that started the process.
    - Your process uses the Is changed operator in a filter condition.

In this case, leave the box unchecked. At the end of the process criteria, a window will appear, as shown in the following screenshot:

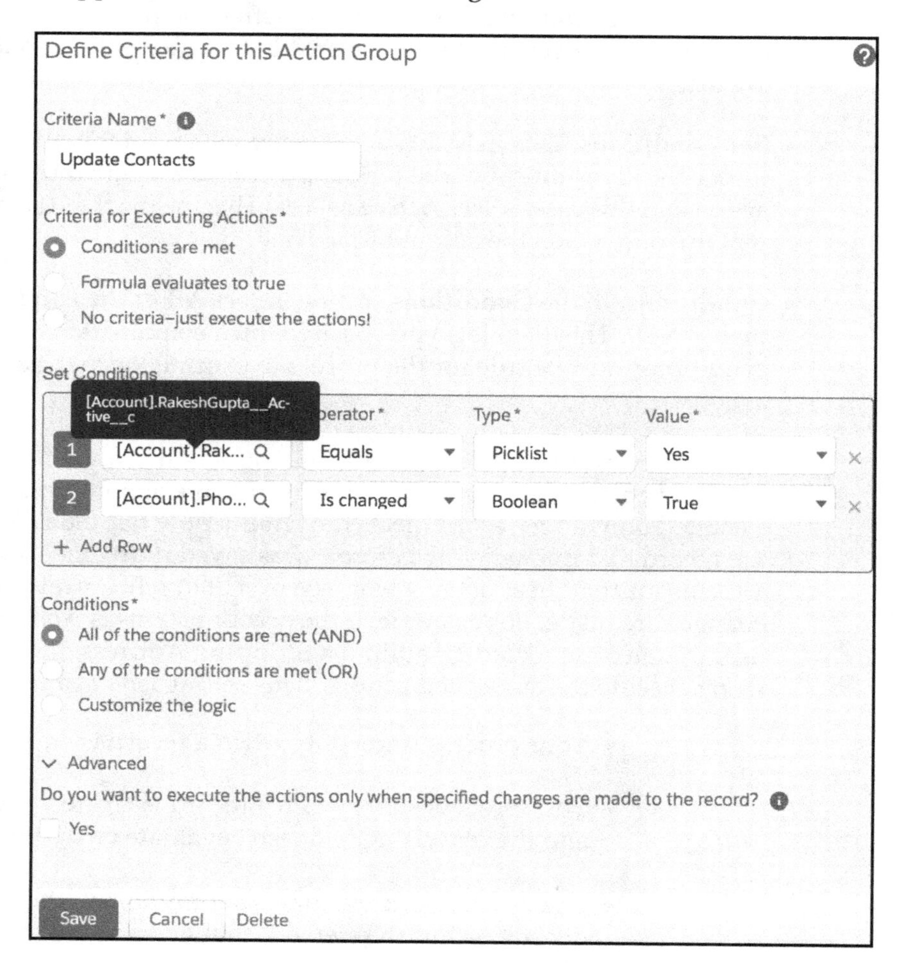

7. Once you are done, click on the **Save** button.

8. Once you are done with the process criteria node, the next step is to add an immediate action to update the related contact's **Asst. Phone** field. For this, we will use the **Update Records** action, available in Process Builder. Click on **Add Action,** available under **IMMEDIATE ACTIONS**. It will open an additional window on the right-hand side of the process canvas screen, where you will have to enter the following details:

- **Action Type**: Select the type of action. In this case, select **Update Records**.
- **Action Name**: Enter a name for the action. Enter `Update Asst. Phone` in **Action Name**.
- **Record Type**: Select the record (or records) that you need to update. Click on the **Record type**; it will pop up a window, where you will see the following options:
    - **Select the Account record that started your process**.
    - **Select a record related to the Account**.
    - These are radio buttons, and only one can be selected to update the child record. In this case, choose **Select a record related to the Account**. Then, in the **Type to filter list**, look for the child object, **Contacts**, as shown in the following screenshot:

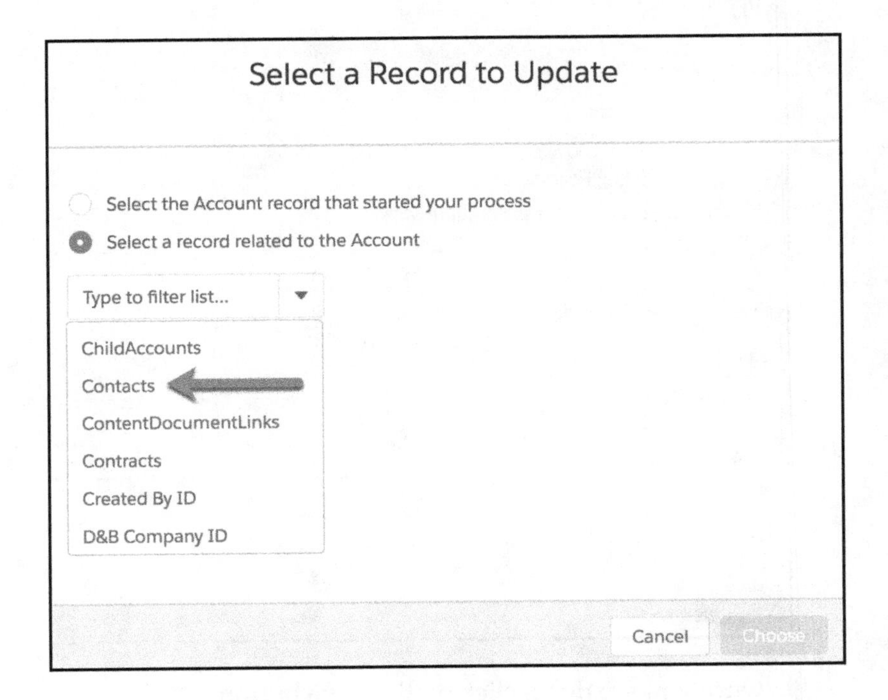

- **Criteria for Updating Records**: Optionally, you can specify conditions to filter the records you are updating. Select **No criteria—just update the records!** There are cool use cases around it, which we will discuss in `Chapter 3`, *Building Efficient and Performance-Optimized Processes.*
- **Field**: Map the `Asst. Phone` field with the `[Account].Phone` field.

Select the **Contacts** object, and then click on the **Choose** button. To select the fields, you can use the **field picker**. To enter the value, use the **text entry** field. It will appear as follows:

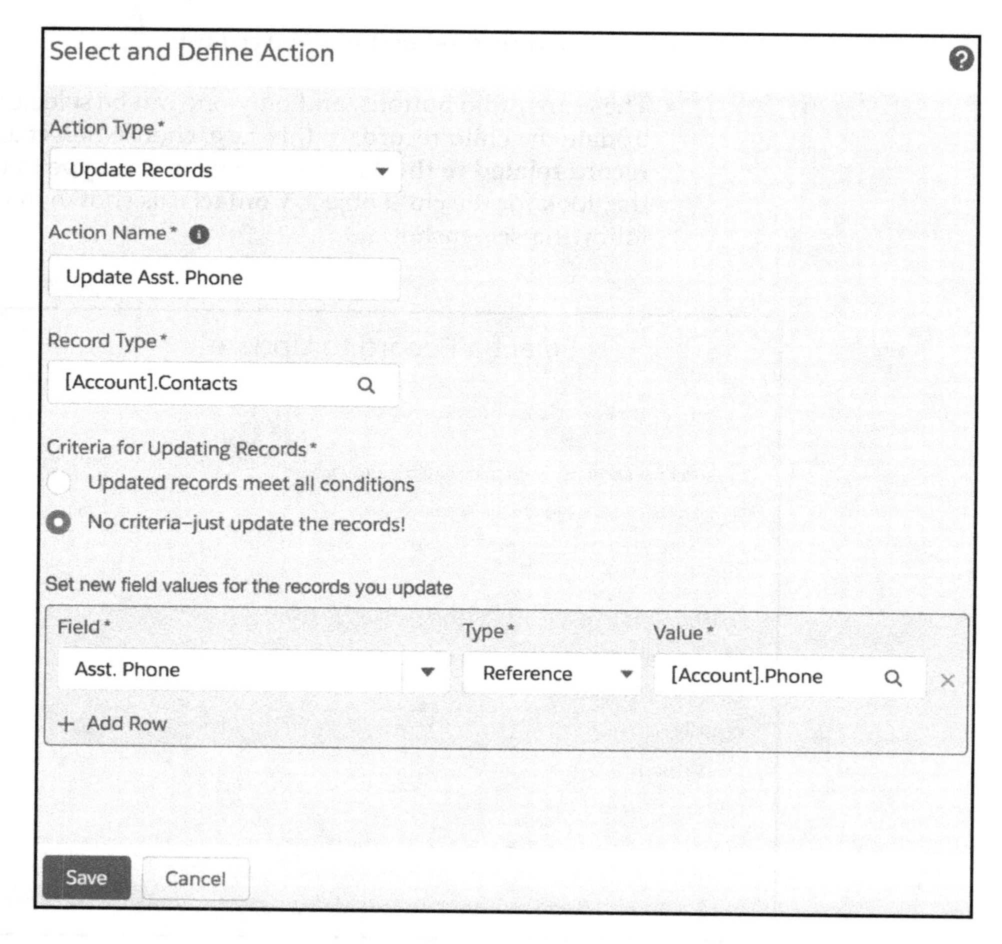

9. Once you are done, click on the **Save** button.

10. The final step is to activate the process. Click on the **Activate** button, available on the button bar. Finally, the process will appear, as shown in the following screenshot:

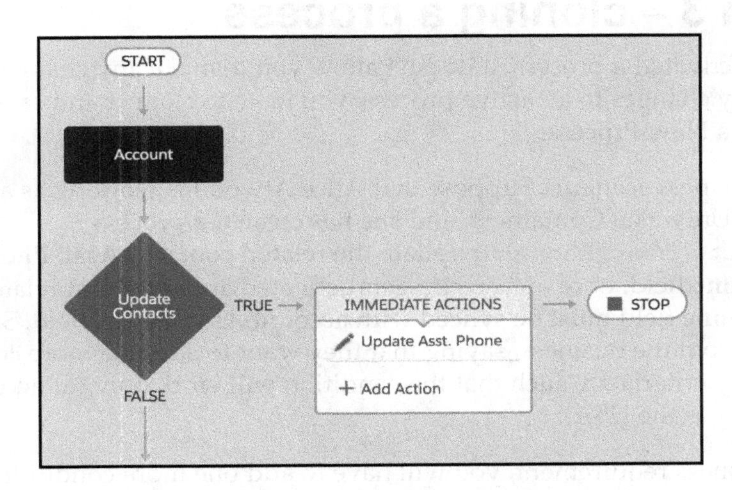

From now on, if users try to update the **Phone** field of an active account, the process will automatically update the related contact's **Asst. Phone** field with the value available in the account's **Phone** field, as shown in the following screenshot:

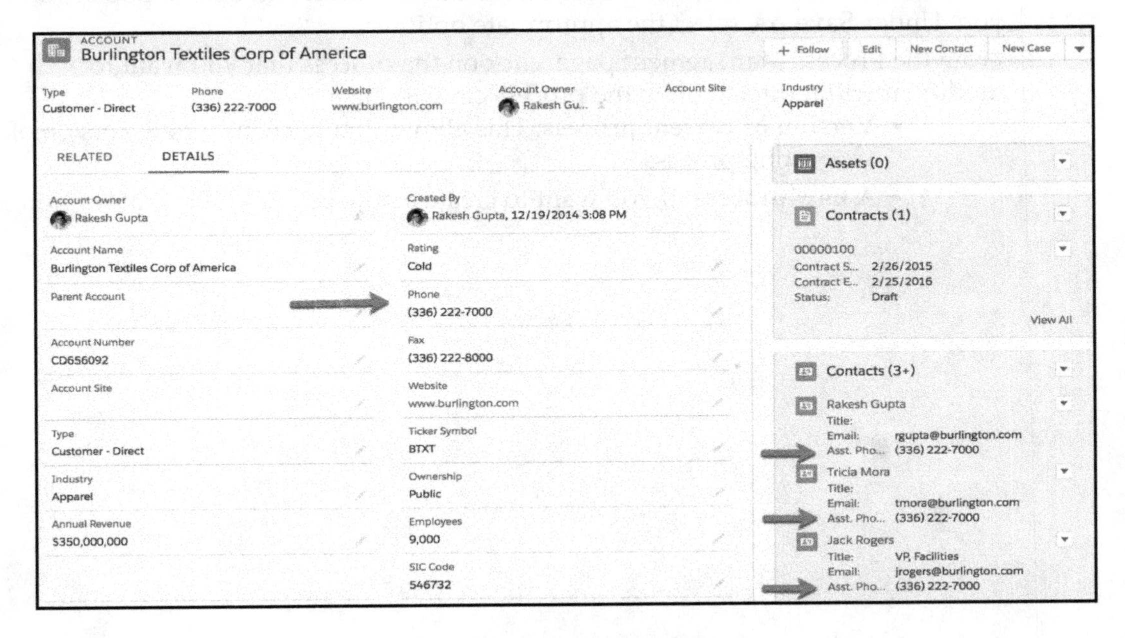

Before testing the process, make sure that you have activated it.

Hands on 3 – cloning a process

Once you have activated a process, it doesn't allow you to make any changes to it. If you want to make any changes to an active process, you have to clone it and save it as either a **New Version** or a **New Process**.

Let's look at a business scenario. Suppose that Alice Atwood is working as a system administrator at Universal Containers, and she has created a process (`Update_Contacts_Asst_Phone`) to update the related contact's **Asst. Phone** field from the account's **Phone** field, once an account gets activated and after that related contacts **Asst. Phone** field must be synced with account **Asst. Phone** field. She has received another request from the business, saying that they want to add one more condition to the process. The entry criteria are such that the condition will work only for accounts wherein the billing country is the USA.

To meet this business requirement, you will have to add one more condition to the existing process. For this, you will have to modify your process. As the process is already activated, the only possibility is to clone it. To clone a process, follow these instructions:

1. Click on the **Clone** button, available on the button bar. It will open a popup for you. Under **Save As**, select the appropriate options:
2. From the Process Management page, click on the process that you want to modify. It will redirect you to the process canvas page:
    - **Version of current process**: This allows you to create a new version of an existing process.
    - **A new process**: If you want to create a new process, select this option.

In this case, select **Version of current process** to create a new version of an existing process, as shown in the following screenshot:

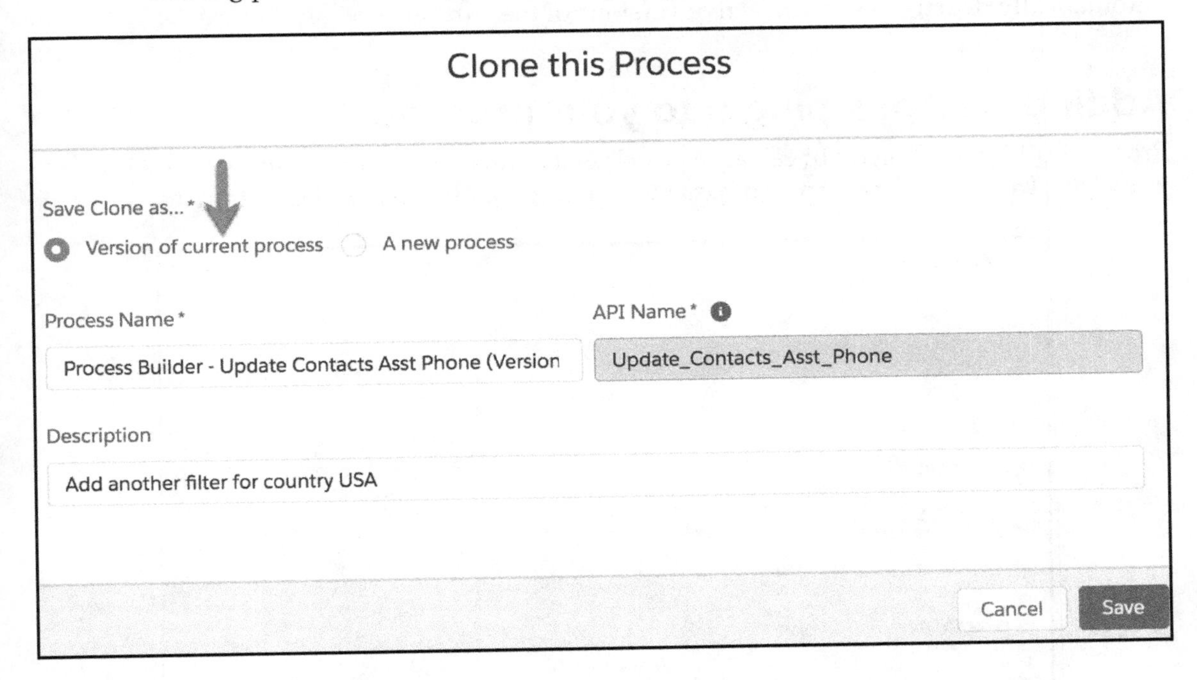

3. Click on the **Save** button when you are done.
4. Now, you can modify the process as per the preceding business requirement, as follows:

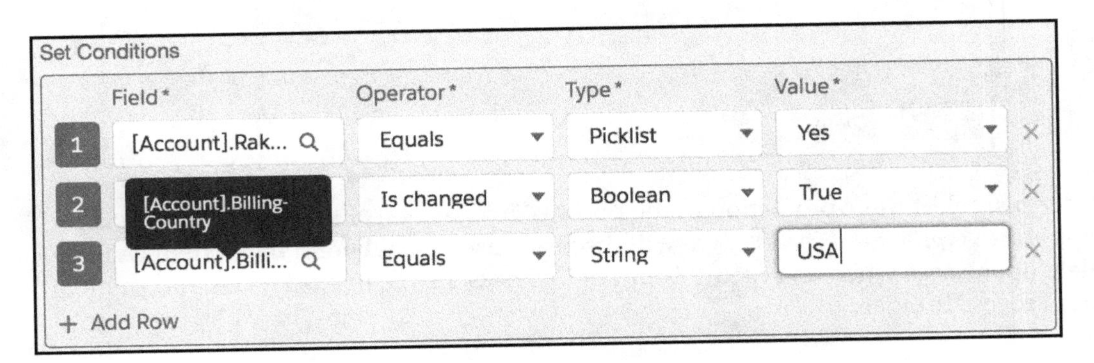

The business requirements will appear, as shown in the preceding screenshot. Once you are done, click on the **Activate** button to activate the current version of the process. Salesforce automatically deactivates others active versions of the current process, if any exist.

Adding an Apex plugin to your process

Process Builder allows you to call an Apex class that includes methods annotated with `@InvocableMethod`. You can also pass the required value in Apex class variables:

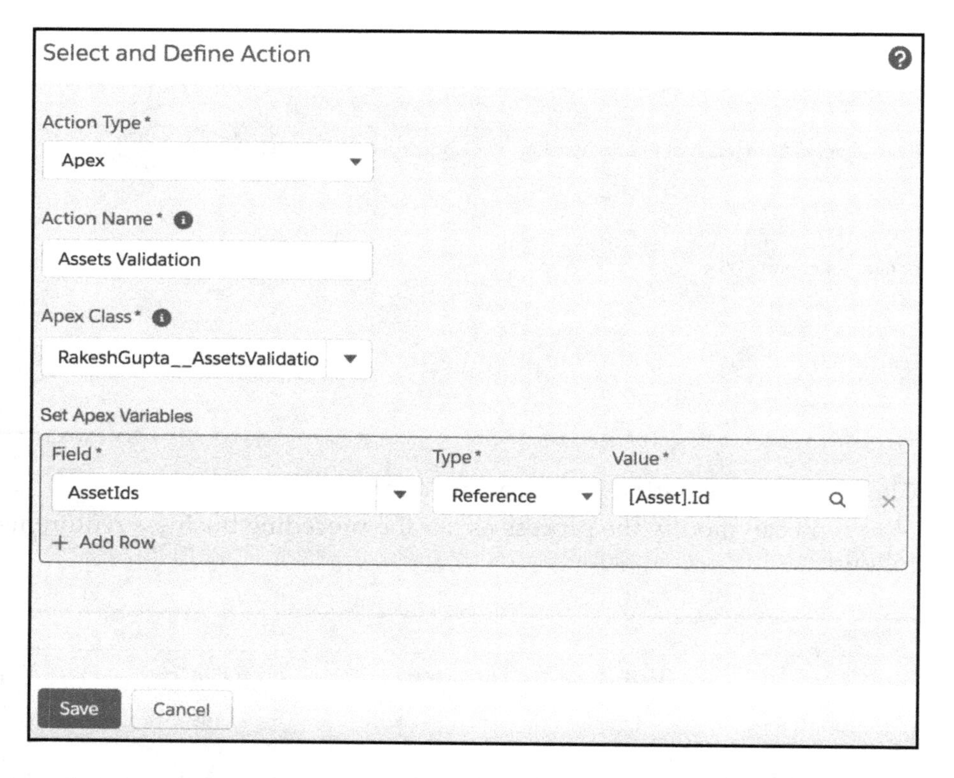

To do this, select the **Apex** action in the process. In `Chapter 3`, *Building Efficient and Performance-Optimized Processes,* we will discuss a use case wherein the system can auto-delete open cases when an account is out of business, using an Apex class and calling it from Process Builder.

Hands on 4 – posting opportunity details to the Chatter group

Often, a business will want to auto-post the details of records to Chatter groups. This is possible through Apex code, but there is a way to achieve it without code – that is, Process Builder. Process Builder has a specific action for this, called **Post to Chatter**. You can auto-post the record details to a Chatter group using the following methods:

- Process Builder
- A combination of Flow and Process Builder
- A Flow and an Inline Visualforce page

Let's look at a business scenario. Suppose that Helina Jolly is working as a system administrator at Universal Containers. She has received a requirement to post opportunity details to the `Sales Executive` Chatter group, whenever an opportunity gets created with an amount greater than $1,000,000.

Follow these instructions to achieve this using Process Builder:

1. Create a public Chatter group, named `Sales Executive`.

Go to `https://help.salesforce.com/HTViewHelpDoc?id=collab_group_creating.htmlanguage=en_US` to learn how to create a Chatter group.

2. To create a process, navigate to **Setup** (gear icon) | **Setup** | **PLATFORM TOOLS** | **Process Automation** | **Process Builder**, click on the **New** button, and enter the following details:
    - **Name**: Enter `Post Opportunity Information to Chatter Group`
    - **API Name**: This will be auto-populated, based on the name
    - **Description**: Write some meaningful text, so that other developers or administrators can easily understand why this process was created
3. Once you are done, click on the **Save** button. This will redirect you to the process canvas, which will allow you to create or modify the process.

4. After defining the process properties, the next task is to select the object upon which you want to create a process and define the evaluation criteria. For this, click on the **Add Object** node. It will open an additional window on the right-hand side of the process canvas screen, where you will have to enter the following details:
    - **Object**: Start typing, and then select the **Opportunity** object.
    - **Start the process**: For **Start the process**, select **only when a record is created**. This means that the process will fire only at the time of record creation.
    - **Recursion – Allow process to evaluate a record multiple times in a single transaction**: Select this checkbox only when you want the process to evaluate the same record up to five times in a single transaction. In this case, leave the box unchecked.
5. Once you are done, click on the **Save** button.
6. After defining the evaluation criteria, the next step is to add the process criteria. Once the process criteria are true, the process will execute the associated actions. To define the process criteria, click on the **Add Criteria** node. It will open an additional window on the right-hand side of the process canvas screen, where you will have to enter the following details:
    - **Criteria Name**: Enter `Only when amount > $1M` in **Criteria Name**.
    - **Criteria for Executing Actions**: Select the type of criteria that you want to define. You can select either **Formula evaluates to true,** or **Conditions are met** (a filter to define the process criteria), or **No criteria-just execute the actions!** In this case, select **Conditions are met**.
    - **Set Conditions**: This field lets you specify which combination of the filter conditions must be true for the process to execute the associated actions. In this case, select a **Value** greater than `$1,000,000`.
    - **Conditions**: In the **Conditions** section, select **All of the conditions are met (AND)**. This field lets you specify which combination of the filter conditions must be true for the process to execute the associated actions.

The dollar sign ($) is not required when entering the amount. What we enter as criteria can be just the raw number – that is, 10,000,000. The system will add the currency for us.

The process criteria will look as follows:

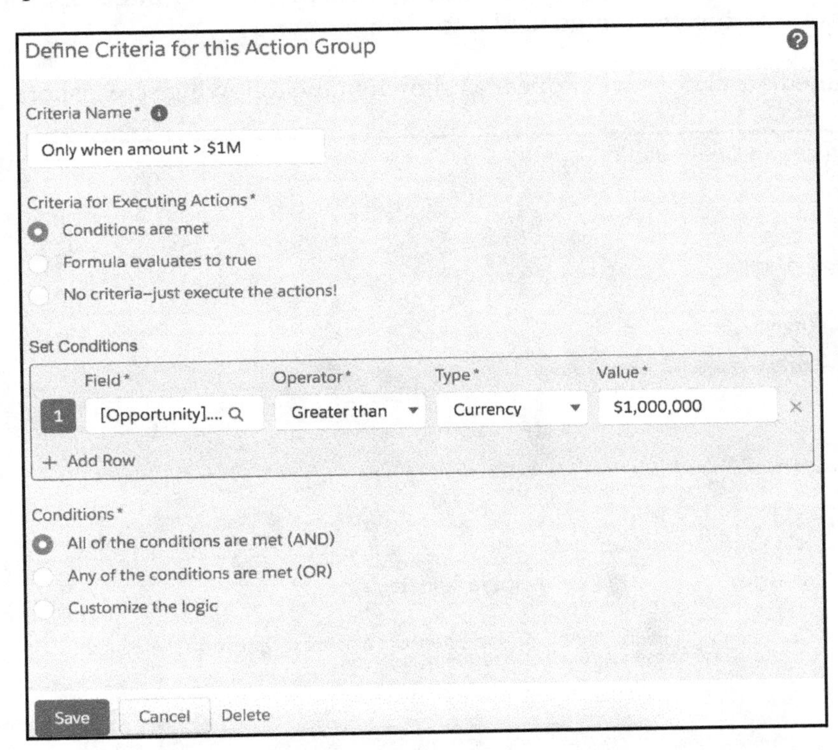

7. Once you are done, click on the **Save** button.
8. Once you are done with the process criteria node, the next step is to add an immediate action to post the opportunity details to the Chatter group. For this, we will use the **Post to Chatter** action, available in Process Builder. Click on **Add Action**, available under **IMMEDIATE ACTIONS**. This will open an additional window on the right-hand side of the process canvas screen, where you will have to enter the following details:
    - **Action Type**: Select the type of action. In this case, select **Post to Chatter**.
    - **Action Name**: Enter `Post to Sales Executive Chatter group` in **Action Name**.

- **Post to**: This allows you to select the **Chatter Group,** or **User,** or **Current Record,** where you want to post the opportunity details. From the drop-down menu, select **Chatter Group**. Then, start typing the name of the Chatter group in the textbox, and select **Sales Executive**.
- **Message**: Enter the message that you want to post. You can also use the fields above the textbox to mention a user or group, add a topic, or insert a merge field into the message.

Immediate actions will appear as shown in the following screenshot:

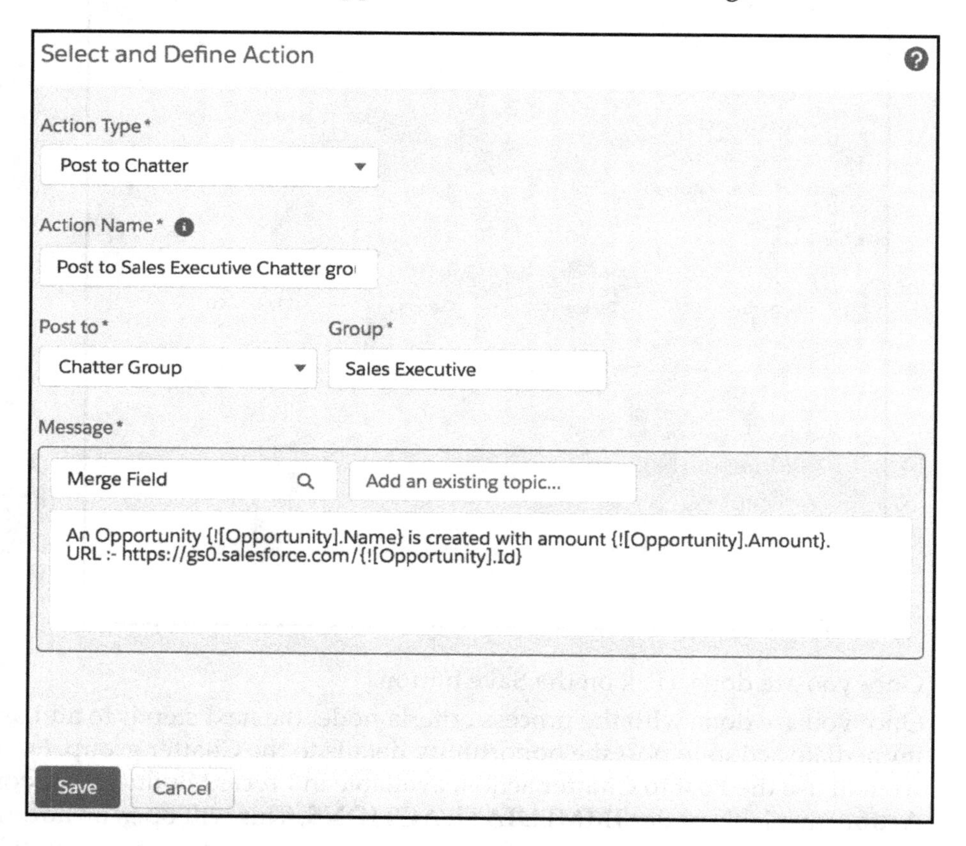

9. To insert fields, click on **Insert Field**. To mention a user or Chatter group, use the **Mention a user or group** textbox. Finally, to add a topic, use the **Add topic** textbox. Once you are done, click on the **Save** button.

10. Once you are done, the final step is to activate the process. Click on the **Activate** button, available on the button bar. Finally, the process will appear, as shown in the following screenshot:

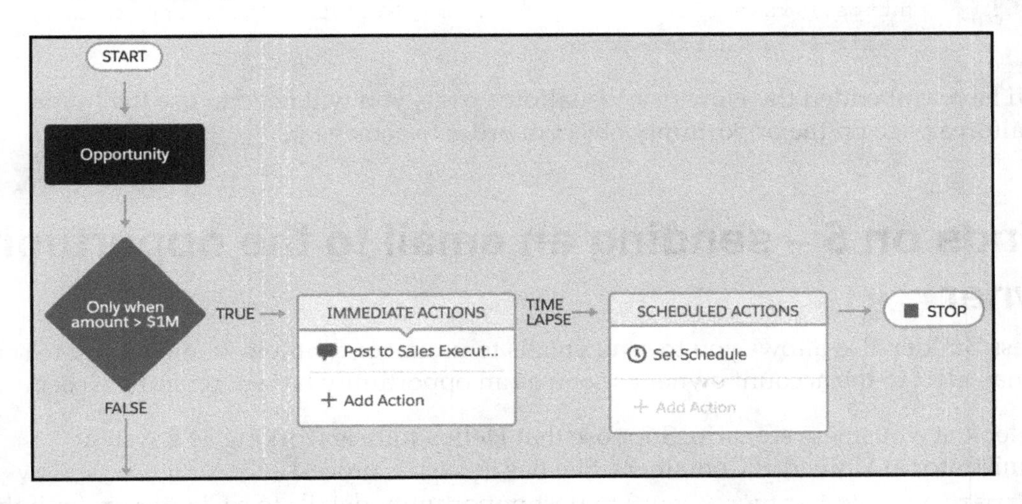

From now on, if users create an opportunity with an amount higher than $1,000,000, the process will auto-post a textpost to the Chatter group `Sales Executive`, which will look as follows:

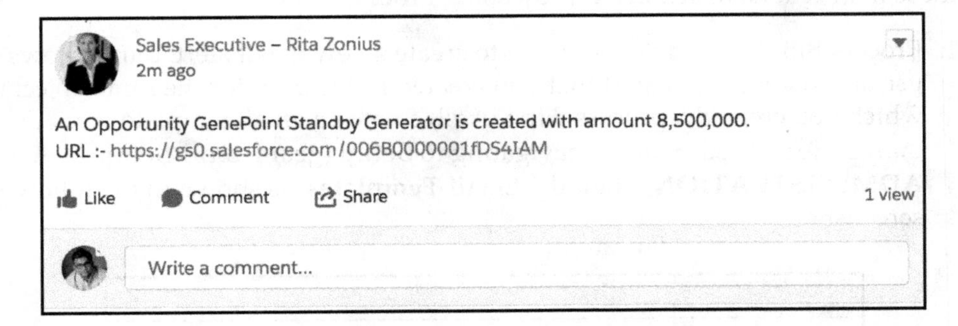

You can also achieve this through a combination of Flow and Process Builder. If you use Process Builder, you can only post a textpost, whereas if you use Flow, you can post a linkpost, textpost, and so on. It will also allow you to set **CreatedById**. For this, use the **Record Create** element and the **FeedItem** object.

You can achieve the preceding requirement using the **Post to Chatter** static action or the **FeedItem** object, available in Flow and Process Builder. To learn more about this, refer to `https://rakeshistom.wordpress.com/2014/06/14/post-opportunity-details-to-a-chatter-group/`.

If you have embedded the Flow in a Visualforce page, you will need to use the Inline Visualforce page on the opportunity object in order to achieve it.

Hands on 5 – sending an email to the opportunity owner

Process Builder also allows you to send emails to users; for example, it allows you to send an email alert to the account owner as soon as an opportunity has successfully closed.

Let's look at a business scenario. Suppose that Helina Jolly is working as a system administrator at Universal Containers. She has created a process (`Post Opportunity Information to Chatter Group`) to post opportunity details to a Chatter group if the amount is greater than $1,000,000. She receives another requirement, to send an email to the opportunity's account owner five days after its creation.

Follow these instructions to achieve this by using Process Builder:

1. Process Builder doesn't allow you to create a new email alert, but it allows you to use an existing email alert that you created in the past, for the same object upon which you created a process. First of all, create an email template, `Account owner notification`, by navigating to **Setup** (gear icon) | **Setup** | **ADMINISTRATION** | **Email** | **Email Templates**, as shown in the following screenshot:

Subject New opportunity gets created for your account - {!Account.Name}

HTML Preview

Dear {!Account.OwnerFullName},

A new opportunity gets created for your account {!Account.Name}. Below is some key information of new opportunity

Name :- {!Opportunity.Name}
Stage:- {!Opportunity.StageName}
Amount:- {!Opportunity.Amount}
Close Date:- {!Opportunity.CloseDate}
Owner:- {!Opportunity.OwnerFullName}

Best Regards,
Universal Containers Sales Team

2. The second step is to create an email alert on the **Opportunity** object by navigating to **Setup** (gear icon) | **Setup** | **PLATFORM TOOLS** | **Process Automation** | **Process Builder** | **Workflow Actions** | **Email Alerts**. Click on the **New Email Alert** button, and save it with the name `Email to Opportunity Owner`. It should look as follows:

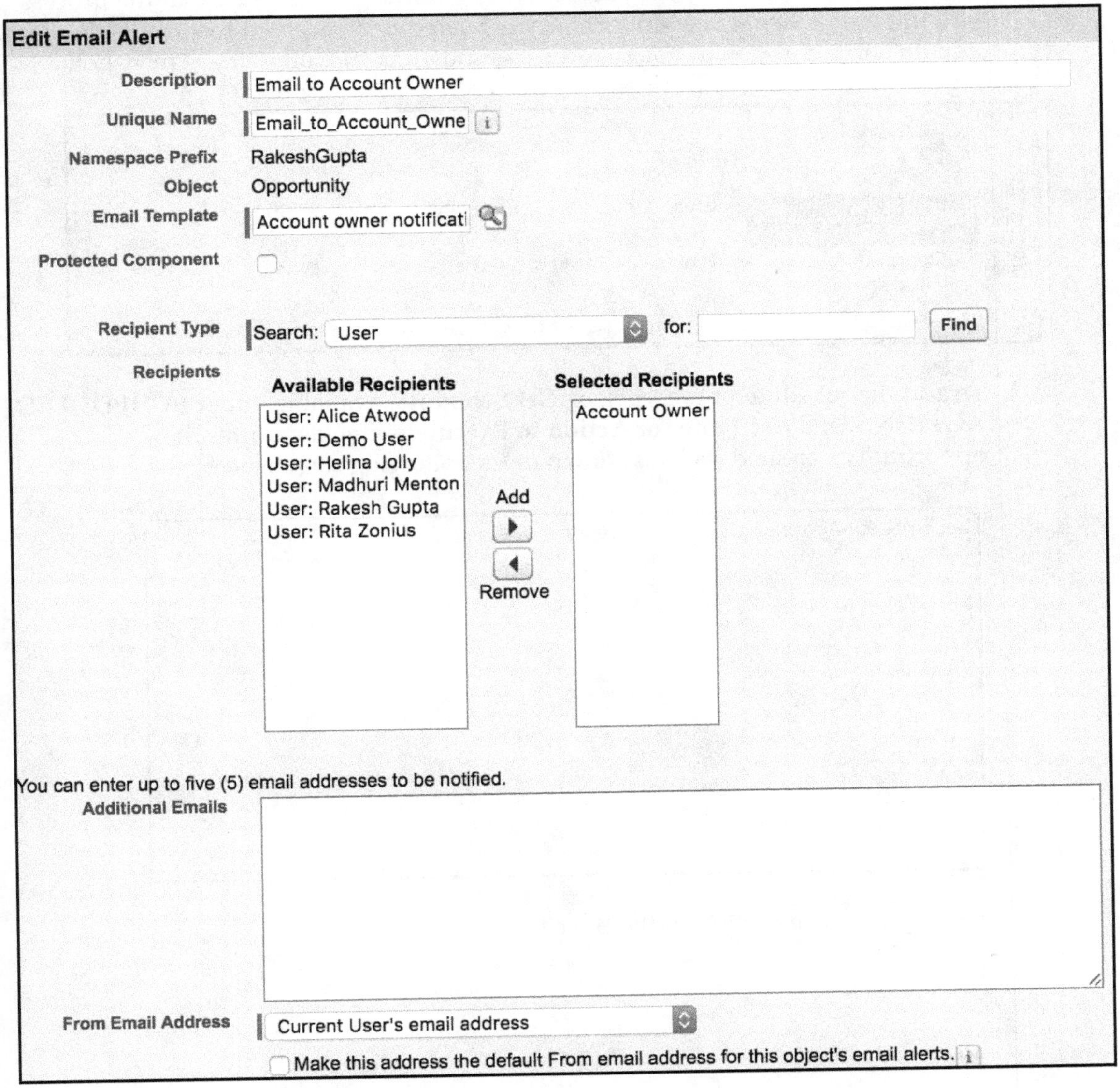

Edit Email Alert

Description: Email to Account Owner
Unique Name: Email_to_Account_Owne
Namespace Prefix: RakeshGupta
Object: Opportunity
Email Template: Account owner notificati
Protected Component
Recipient Type: Search: User for: Find
Recipients

Available Recipients:
User: Alice Atwood
User: Demo User
User: Helina Jolly
User: Madhuri Menton
User: Rakesh Gupta
User: Rita Zonius

Add
Remove

Selected Recipients:
Account Owner

You can enter up to five (5) email addresses to be notified.
Additional Emails
From Email Address: Current User's email address
Make this address the default From email address for this object's email alerts.

3. To add this email alert to the existing process, navigate to **Setup** (gear icon) | **Setup** | **PLATFORM TOOLS** | **Process Automation** | **Process Builder**. Open the `Post Opportunity Information to Chatter Group` process that you created to post the opportunity details to the Chatter group. Save it as a **New Version,** because you can't modify an activated process.
4. Now, we have to modify the process entry criteria, to make sure that the process will fire only when a new opportunity is related to an account record. To do that, add an entry criterion in your process, as shown in the following screenshot:

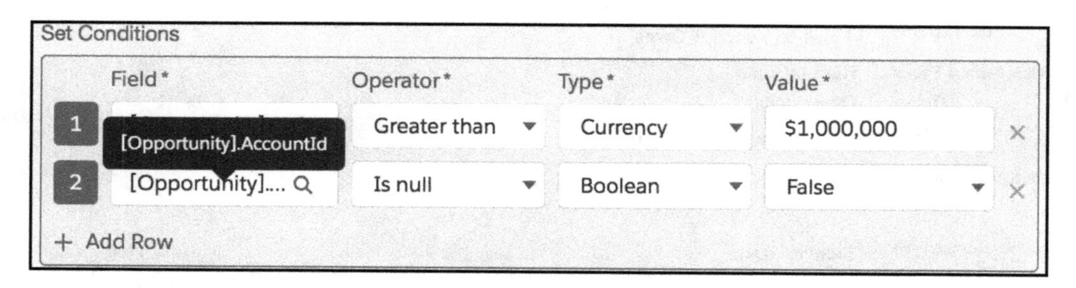

5. To add the scheduled time, click on **Set Schedule,** available under **SCHEDULED ACTIONS**, and **Set Time for Action to Execute** to five days after the opportunity's created date, as shown in the following screenshot:

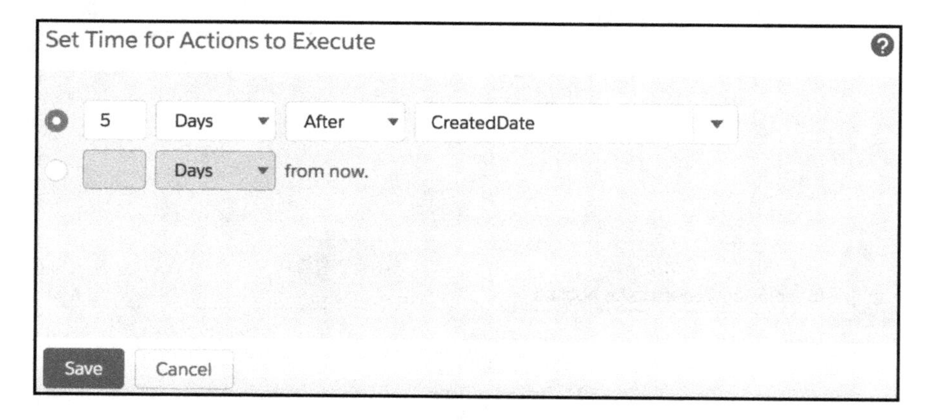

6. Once you are done, click on the **Save** button.

7. The next step is to add a scheduled action to send an email. For this, we will use the **Send an Email** action, available in Process Builder. To add schedule actions, click on **Add Action,** available under **SCHEDULED ACTIONS**. This will open an additional window on the right-hand side of the process canvas screen, where you will have to enter the following details:
    - **Action Type**: Select the type of action. In this case, select **Email Alerts**.
    - **Action Name**: Enter `Email to Account Owner`.
    - **Email Alert**: Select the existing email alert. In this case, select the email alert, **Email to Account Owner,** which you created in step 2.

   Scheduled actions will appear, as shown in the following screenshot:

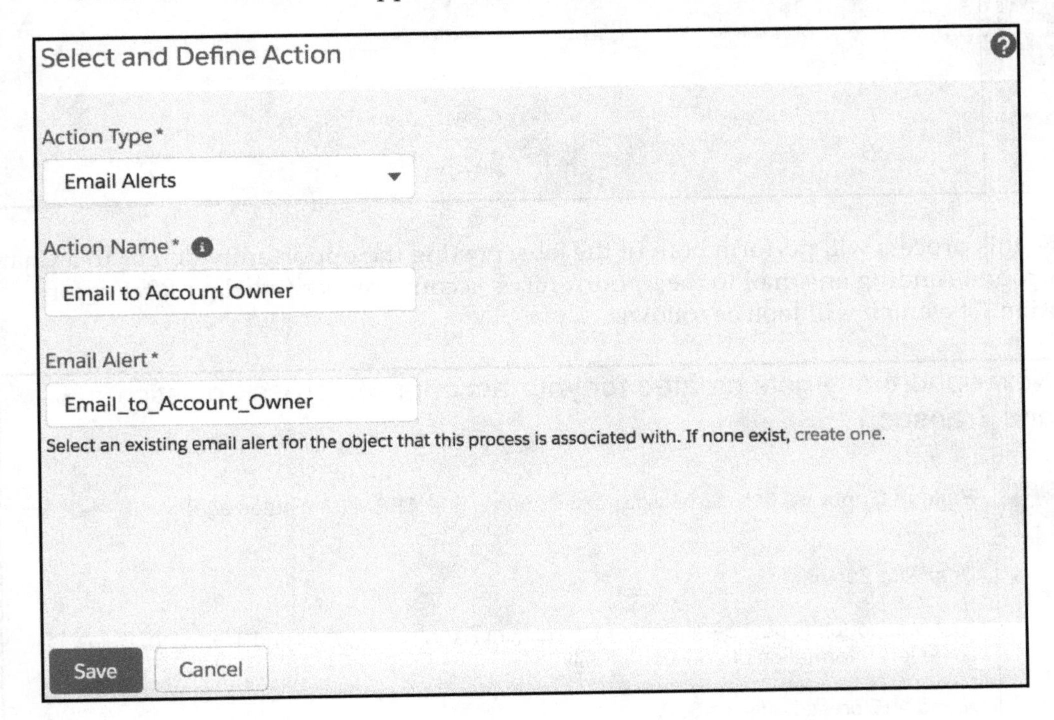

8. Once you are done, click on the **Save** button.

9. Once you are done with the process creation, don't forget to activate the process by clicking on the **Activate** button. Finally, the process will appear, as shown in the following screenshot:

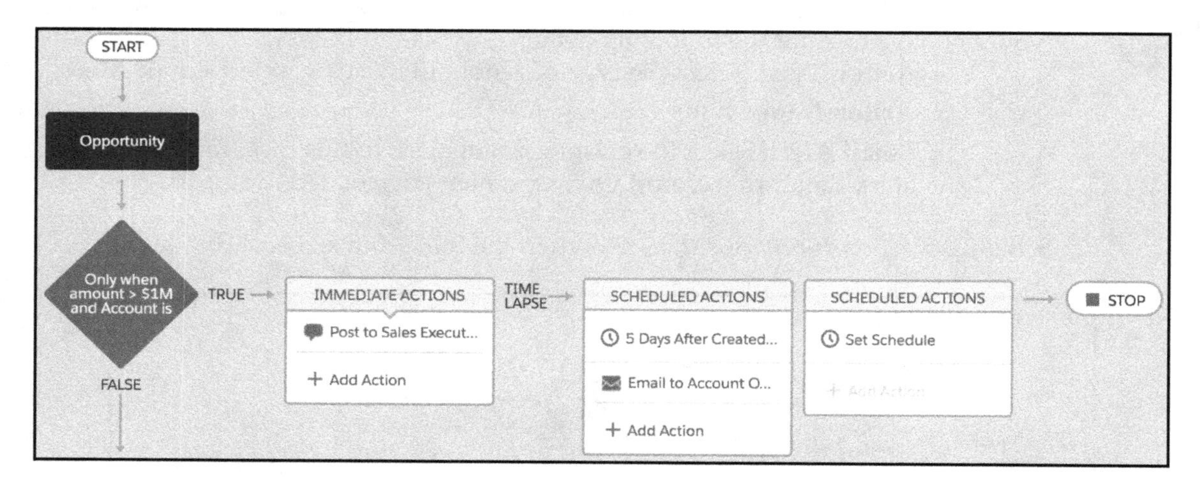

Now, this process will perform both of the jobs: posting the opportunity details to a Chatter group, and sending an email to the opportunity's account owner five days after record creation. The email will look as follows:

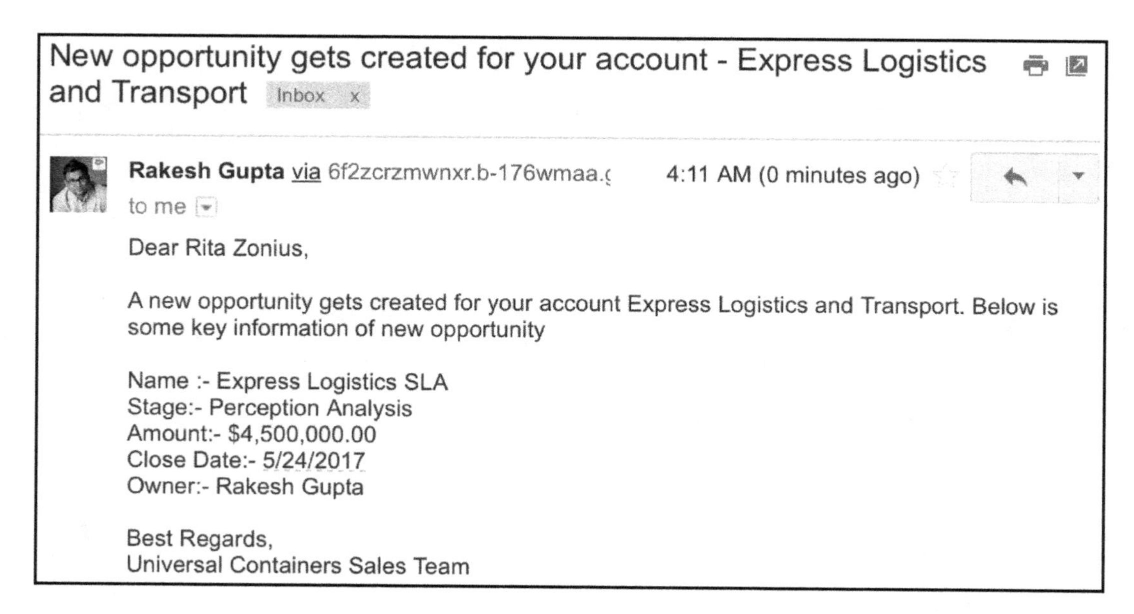
New opportunity gets created for your account - Express Logistics and Transport Inbox x

Rakesh Gupta via 6f2zcrzmwnxr.b-176wmaa.(4:11 AM (0 minutes ago)
to me

Dear Rita Zonius,

A new opportunity gets created for your account Express Logistics and Transport. Below is some key information of new opportunity

Name :- Express Logistics SLA
Stage:- Perception Analysis
Amount:- $4,500,000.00
Close Date:- 5/24/2017
Owner:- Rakesh Gupta

Best Regards,
Universal Containers Sales Team

Flow also allows you to send emails. You have to use the **Send Email** static action, available inside of Flow. To learn more about this, go to `https://rakeshistom.wordpress.com/2014/09/08/reminder-email-to-upload-chatter-profile-photo/`.

If you have embedded the Flow in a Visualforce page, you will need to use an Inline Visualforce page on the opportunity object to fulfill the preceding business use case.

Hands on 6 – checking time-dependent actions from Process Builder

To monitor a time-dependent action queue in Process Builder, follow these instructions:

1. Navigate to **Setup** (gear icon) | **Setup** | **PLATFORM TOOLS** | **Process Automation** | **Flows**.
2. Go to the **Paused and Waiting Interviews** section; from there, you can check out the time-dependent action queue in Process Builder, as shown in the following screenshot:

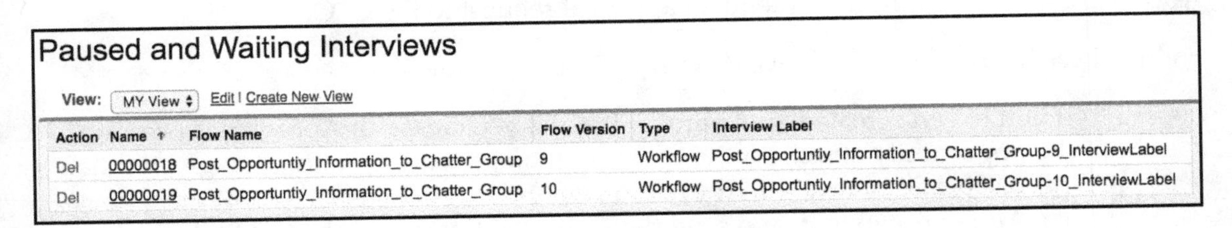

Paused and Waiting Interviews

View: MY View | Edit | Create New View

Action	Name ↑	Flow Name	Flow Version	Type	Interview Label
Del	00000018	Post_Opportuntiy_Information_to_Chatter_Group	9	Workflow	Post_Opportuntiy_Information_to_Chatter_Group-9_InterviewLabel
Del	00000019	Post_Opportuntiy_Information_to_Chatter_Group	10	Workflow	Post_Opportuntiy_Information_to_Chatter_Group-10_InterviewLabel

3. To remove a record from the time-based queue, use the **Del** link, as shown in the preceding screenshot.

This is because Process Builder uses a Flow **Wait** element to create a time-dependent action.

Hands on 7 – submitting a record to an Approval Process

Process Builder also allows you to auto-submit a record to an Approval Process. Currently, the user has to manually submit a record for the Approval Process. You can achieve this type of requirement through the following methods:

- Using Process Builder.
- Using a combination of Flow (submit for the approval-static action) and Process Builder.
- Using a combination of Flow and the Inline Visualforce page on the object's detail page.

Let's look at a business scenario. Suppose that Helina Jolly is working as a system administrator at Universal Containers. She has created a process (`Post Opportunity Information to Chatter Group`) to post opportunity details to a Chatter group, if the amount is greater than $1,000,000, and to send an email to the opportunity's owner. She receives another requirement: to auto-submit new opportunity records to the Approval Process if the amount is greater than $1,000,000. She has already created a one-step Approval Process for this. It will send an approval request to the CEO.

Follow these instructions to achieve this using Process Builder:

1. If you haven't created an Approval Process yet, create an Approval Process on the opportunity object, set the entry criteria opportunity amount to greater than 1,000,000, and save it with the name, `Opportunity amount greater than 1M`. Add one step to it, and send the Approval Request to the CEO role. Make sure that you have activated the Approval Process:

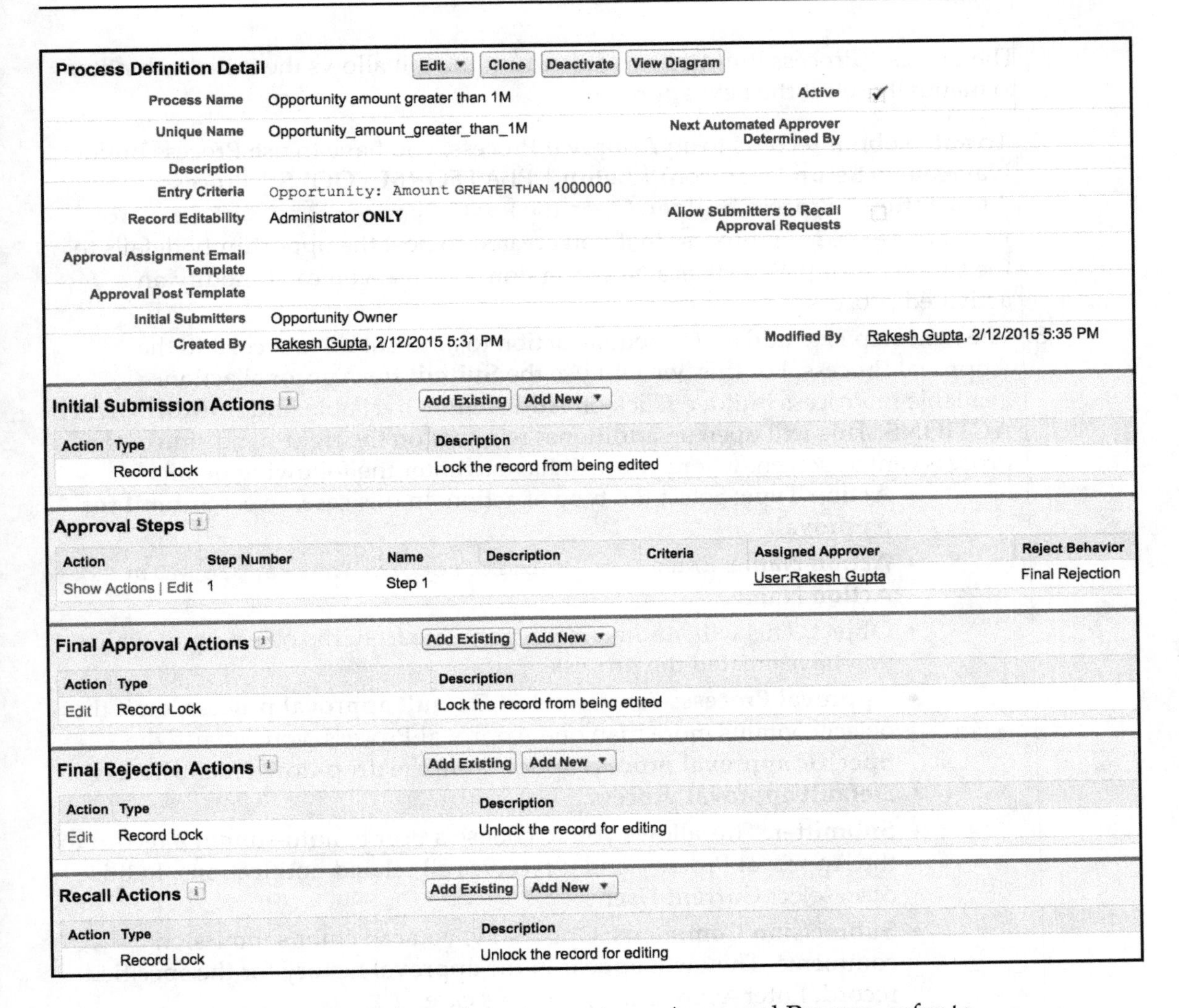

If you want to learn how to create an Approval Process, refer to `https://help.salesforce.com/HTViewHelpDoc?id=approvals_creating_approval_processes.htm&language=en_US`.

The process (Process Builder) will fail at runtime if it allows the initial submitter to manually select the next approver.

2. To auto-submit records to an Approval Process, you have to use Process Builder. Navigate to **Setup** (gear icon) | **Setup** | **PLATFORM TOOLS** | **Process Automation** | **Process Builder**. Open the `Post Opportunity Information to Chatter Group` process that you created to post the opportunity details to the Chatter group. Save it as a **New Version,** because you can't modify an activated process.
3. The next step is to add an immediate action to auto-submit a record to the Approval Process. For this, we will use the **Submit for Approval** action, available in Process Builder. Click on **Add Action,** available under **IMMEDIATE ACTIONS**. This will open an additional window on the right-hand side of the process canvas screen, where you will have to enter the following details:
    - **Action Type**: Select the type of action. In this case, select **Submit for Approval**.
    - **Action Name**: Enter `Auto submit record into approval` in **Action Name**.
    - **Object**: This will automatically populate from the object upon which you have created the process.
    - **Approval Process**: You can select **Default approval process**, or, if the object contains more than one Approval Process, you can use the **Specific approval process** option from the drop-down menu. Select **Default approval process**.
    - **Submitter**: This allows you to choose a user to auto-submit a record to the Approval Process, and it receives all related notifications. In this case, select **Current User**.
    - **Submission Comments**: Optionally, you can enter submission comments. They will appear in the approval history for the specified record. Enter `Auto submit`, in this case.

The immediate actions will look as follows:

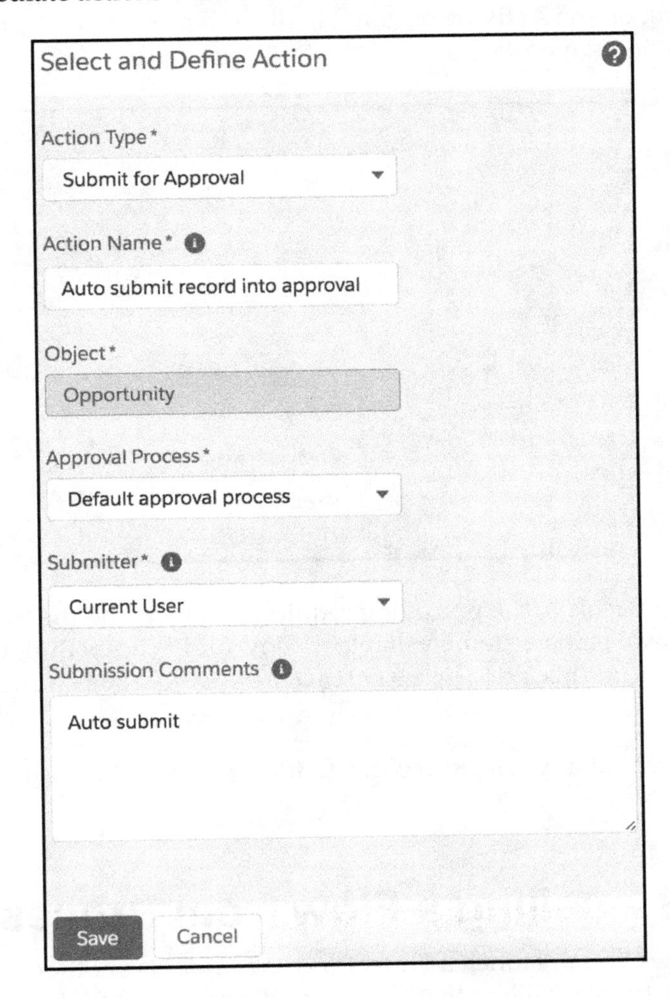

4. Once you are done, click on the **Save** button.

5. Once you are done with the process creation, don't forget to activate the process by clicking on the **Activate** button. Finally, the process will appear, as shown in the following screenshot:

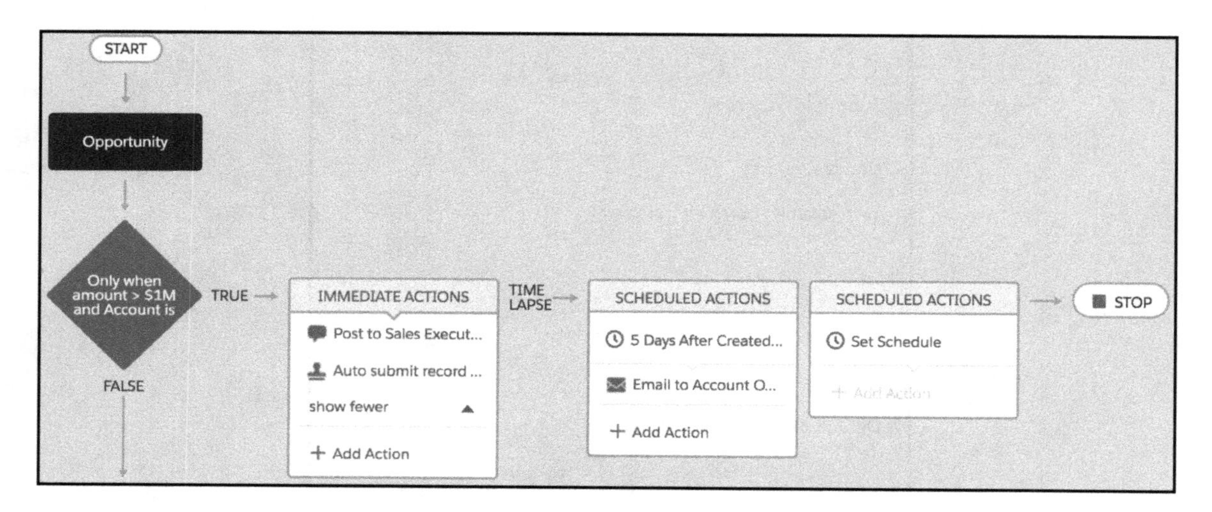

You can achieve the preceding requirement by using the **Submit for Approval** static action, available in Flow and Process Builder. To learn more about this, go to `https://rakeshistom.wordpress.com/2014/06/27/auto-submit-record-into-approval-process-with-flow/`.

It is not possible to submit any related records to the Approval Process using Process Builder.

Hands on 8 – calling a Flow from Process Builder

Process Builder allows you to launch a Flow. Process Builder is another way to auto-launch a Flow. For example, there is a Flow that allows for the removal of followers from closed opportunity records. If you want this Flow to automatically execute whenever the opportunity status gets closed, you should use Process Builder.

Some Flows don't require any user interaction to start; for example, a Flow of the type **Autolaunched Flow**. An Autolaunched Flow can be launched without user interaction; from Process Builder or the Apex `interview.start` method, for example.

Let's look at a business scenario. Suppose that Sara Bareilles, who is working as a system administrator at Universal Containers, has created a Flow to create a new opportunity using the **Record Create** element, as shown in the following screenshot:

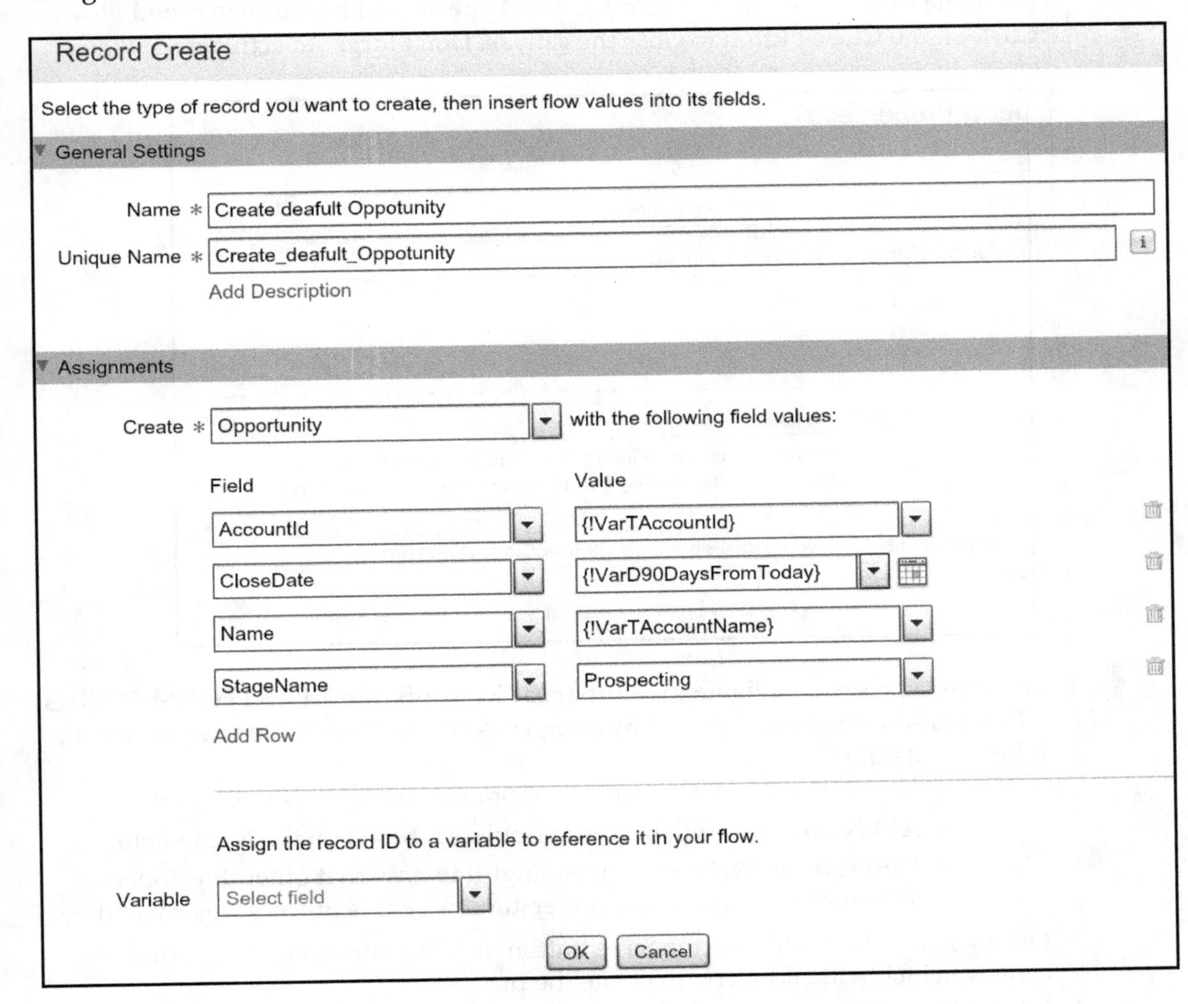

Currently, she is using a custom button on the Account Page layout to call the Flow. She wants to use Process Builder to automatically fire the Flow whenever an account gets created. The `VarD90DaysFromToday` object is a date variable, and the `VarTAccountId` and `VarTAccountName` variables used in the preceding example are nothing but text variables. We will use Process Builder to pass the values in these variables.

Follow these instructions to call a Flow from the process:

1. Create a Flow, similar to what is shown in the preceding screenshot; save it with the name `Create an Opportunity`. The **Type** should be **Autolaunched Flow**. Click on the **Close** button to close the canvas. Don't forget to activate the Flow:

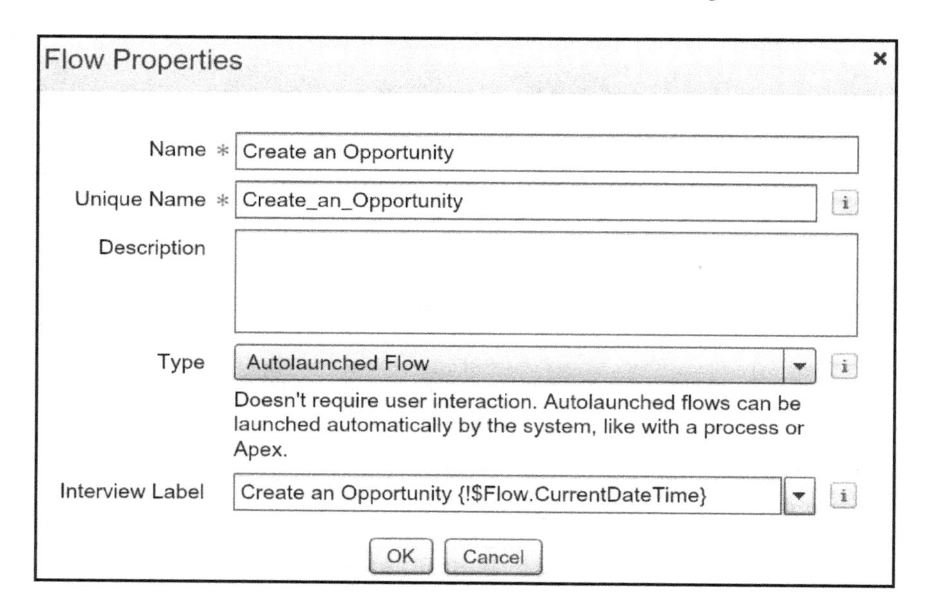

2. To create a process, navigate to **Setup** (gear icon) | **Setup** | **PLATFORM TOOLS** | **Process Automation** | **Process Builder**, click on the **New Button**, and enter the following details:
    - **Name**: Enter `Auto create an Opportunity` in **Name.**
    - **API Name**: This will be auto-populated, based on the **Name** field.
    - **Description**: Write some meaningful text, so that other developers or administrators can easily understand why this process was created.
3. Once you are done, click on the **Save** button. It will redirect you to the process canvas, which will allow you to create the process.
4. After defining the process properties, the next task is to select the object upon which you want to create a process and define the evaluation criteria. For this, click on the **Add Object** node. This will open an additional window on the right-hand side of the process canvas screen, where you will have to enter the following details:
    - **Object**: Start typing the name, and then select the **Account** object.

- **Start the process**: For **Start the process**, select **only when a record is created**. This means that the process will fire only at the time of record creation.
- **Recursion: Allow process to evaluate a record multiple times in a single transaction?**: Select this checkbox only when you want the process to evaluate the same record up to five times in a single transaction. In this case, leave the box unchecked.

4. Once you are done, click on the **Save** button.
5. After defining the evaluation criteria, the next step is to add the process criteria. To define the process criteria, click on the **Add Criteria** node. This will open an additional window on the right-hand side of the process canvas screen, where you will have to enter the following details:
   - **Criteria Name**: Enter `Always` as the criteria name.
   - **Criteria for Executing Actions**: Select the type of criteria that you want to define. You can select either **Formula evaluates to true,** or **Conditions are met** (a filter to define the process criteria), or **No criteria-just execute the actions!** In this case, select **No criteria-just execute the actions!** This means that the process will fire in every condition.
6. Click on the **Save** button.
7. Once you are done with the **process criteria** node, the next step is to add an immediate action to launch a Flow. For this, we will use the **Flows** action, available in Process Builder. Click on **Add Action,** available under **IMMEDIATE ACTIONS**. This will open an additional window on the right-hand side of the process canvas screen, where you will have to enter the following details:
   - **Action Type**: Select the type of action. In this case, select **Flows.**
   - **Action Name**: Enter `Auto create new Opportunity` in **Action Name.**
   - **Flow**: Select the Flow that you want to execute. In this case, select the **Create an Opportunity** Flow.
   - **Set Flow Variables**: Use this to pass the value in your Flow variables. For the current use case, map the `VarD90DaysFromToday` variable with the formula `TODAY() + 90`, **the** `VarTAccountId` variable with `[Account].Id`, and the `VarTAccountName` variable with `[Account].Name`.

The immediate actions will look as follows:

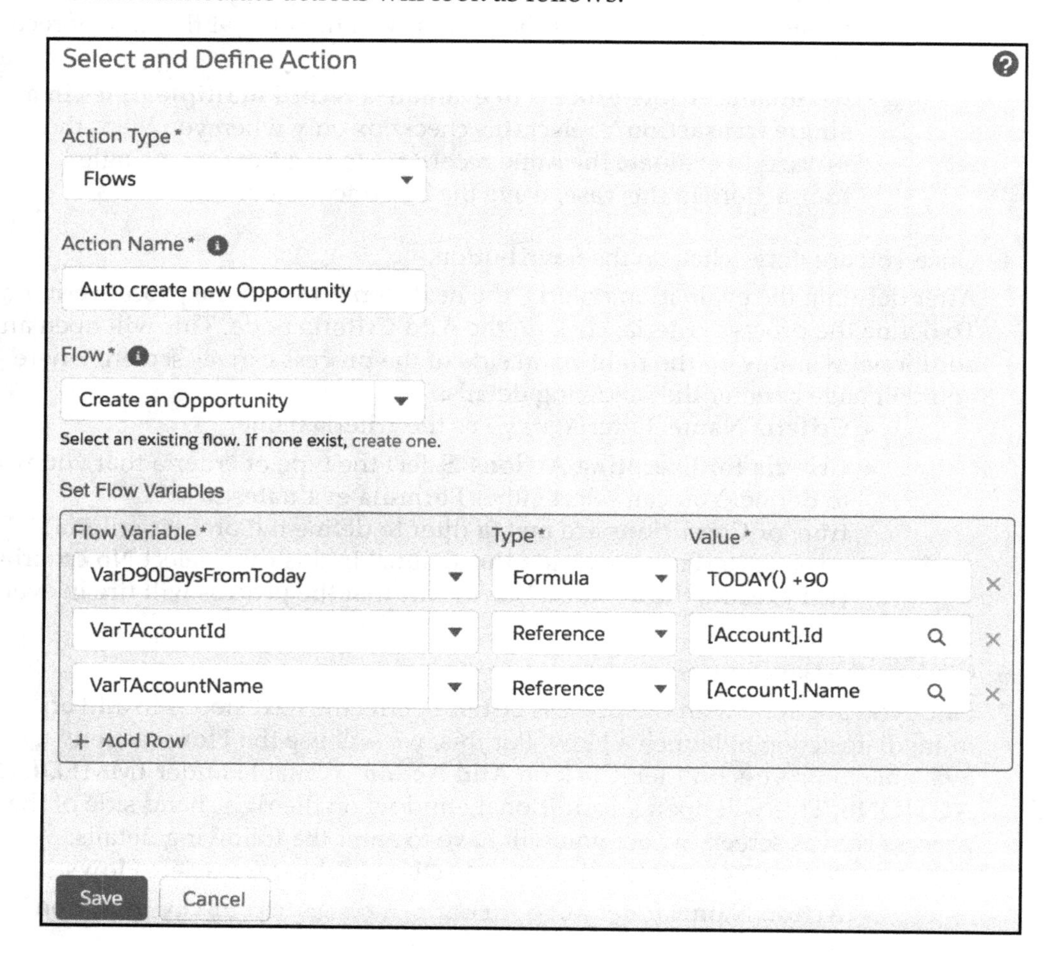

9. To assign the value to multiple variables, click on the **Add Row** link. Once you are done, click on the **Save** button.
10. The final step is to activate the process. Click on the **Activate** button, available on the button bar. Finally, the process will appear, as shown in the following screenshot:

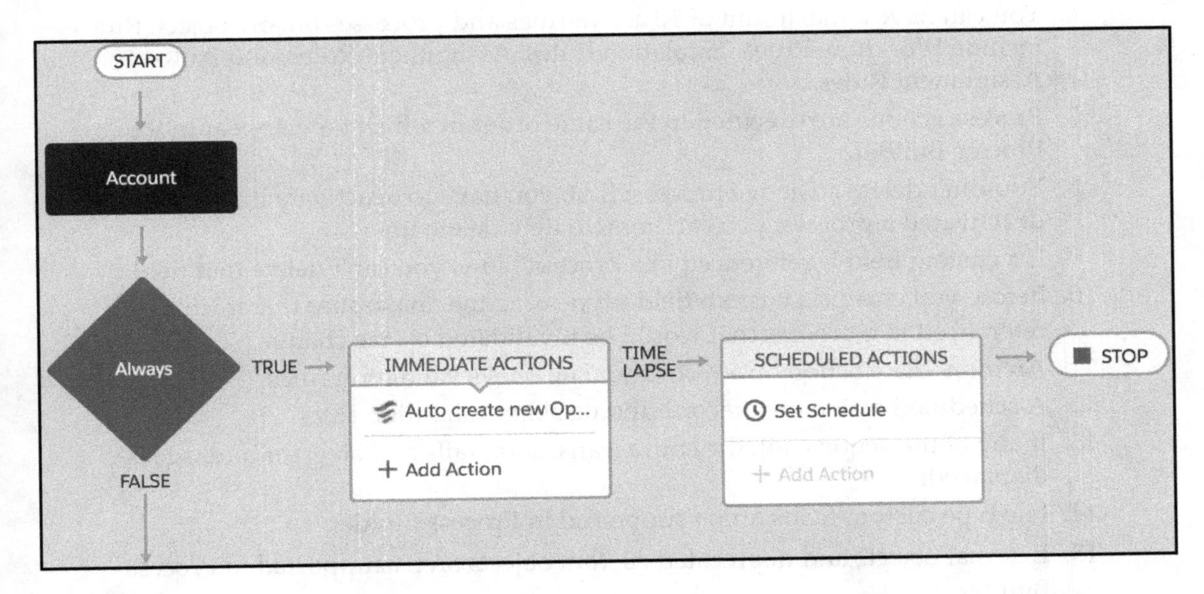

From now on, if an account gets created, Flow will automatically be executed by Process Builder, and a new opportunity will be created.

A few points to remember

The following are some noteworthy points regarding Process Builder:

1. A process's API name must be less than, or equal to, 77 characters, and it's always associated with a single object.
2. Using Process Builder, you can't delete records. If you want to do that, you have to use Flow with Process Builder.
3. To set the **Text** data field to blank, you can use `{!$GlobalConstant.EmptyString}`.

4. By default, the process owner will receive an email from Salesforce. If the process fails at runtime, or if any fault occurs, the error or warning messages might refer to a Flow instead of a process. If you want to notify other users and your IT department, use the **Apex Exception Email**, under the Salesforce setup.
5. The process follows all of the governor limits that apply to Apex.
6. You can have a maximum of 50 active rules and processes on any object. Rules include Workflow Rules, Escalation Rules, Assignment Rules, and Auto-Assignment Rules.
7. Process actions are executed in the same order in which they appear in the Process Builder.
8. You can't delete an active process; first, you have to deactivate it. Once you have deactivated a process, you can immediately delete it.
9. If a custom field is referenced in a process/Flow, you can't delete that field.
10. Before you change a custom field's type or name, make sure that it isn't referenced in a process that would be invalidated by the change.
11. An immediate action on a record update obeys validation rules.
12. A scheduled action on a record update skips validation rules.
13. If any of the actions fail, the entire transaction fails and an error message is displayed.
14. File type custom fields aren't supported in Process Builder.
15. External objects and deprecated custom objects aren't supported in Process Builder.
16. If a single action group includes multiple **Update Records** actions that apply different values to the same field, then the last action's value is used.
17. The total number of process criteria nodes that are evaluated and actions that are executed at runtime is 2,000.

Exercises

1. Create a one-step Approval Process for an account, to get an approval from the CEO once an account becomes active. Use Process Builder to automatically submit an account for approval, once it gets activated.
2. Automatically update the lead rating to **Warm** whenever the lead status gets updated to **Working—Contacted**.

3. Develop an application using Flow and Process Builder. It will automatically remove all of the followers from an opportunity, except for the opportunity owner, once its stage is updated to **Closed-Won** or **Closed-Lost**.
4. Automatically create a child case whenever a case is created with a **High** priority. Use the following information to create the child case:
    - **Status**: New
    - **Priority**: Medium
    - **Case Origin**: Phone
    - **Associate it with a high priority case**
5. Create a one-step Approval Process for an account, to get an approval from the CEO once an account becomes active. Use Process Builder to automatically submit an account for approval once it gets activated.
6. Develop an application, using Flow and Process Builder, that will automatically add a new user that has the role of Sales Executive to the `Sales Executive` Chatter group.

If the `Sales Executive` Chatter group does not exist in your Salesforce organization, create a new one.

7. Once the account has been created, auto-post the account details to a Chatter group (`Sales Executive`).
8. Once the opportunity has successfully closed, auto-post a linkpost on a Chatter group (`Sales Executive`). The output should look similar to the following screenshot:

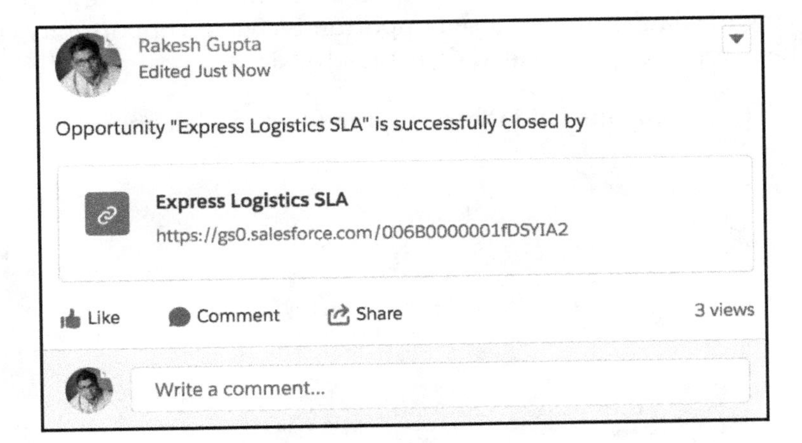

9. Auto-create a child case whenever the case's **Type** field changes to `Mechanical`, and assign it to the `Mechanical Engineers` queue.

If the `Mechanical Engineers` queue does not exist in your Salesforce organization, create a new one.

10. Whenever a case is created, automatically add the user Helina Jolly to the case team members.

Create a new user in your organization, named Helina Jolly.

11. First, set the OWD for Lead to **Private**. Now, develop an application, using Flow and Process Builder, that will automatically share new lead records with the user Helina Jolly, and grant her read/write access.
12. Once an account becomes inactive, send an email notification to all opportunity owners related to that account. Also, update any related open opportunities' statuses to **Closed Lost**.

Use the **Active** field to decide the account status.

13. Once an opportunity has successfully closed, auto-create a new opportunity by copying the data from the closed opportunity. Set the opportunity closed date as the new opportunity's created date, plus 100 days.

Copy only the field that is required to create an opportunity.

14. Create a process that will automatically add new users that have the profile `APAC Support Agent` to a queue (`High Priority Accounts`).

If the `High Priority Accounts` queue and the `APAC Support Agent` profile do not exist in your Salesforce organization, create new ones.

Summary

In this chapter, we went over the various actions available in Process Builder. We looked at a way to auto-create and update a record. We also covered various ways to call a Flow from Process Builder. Then, we moved on to discussing ways to auto-submit a record to the Approval Process and send emails to users. We also discussed how to call Apex from Process Builder. Finally, we discussed the key points of Process Builder. In the next chapter, we will discuss concepts related to deploying and distributing Flows and Processes. We will also look at concepts such as how to debug your Flow and send a process error email to users other the process owner

2
Deploying, Distributing, and Debugging Processes

In the previous chapter, we took a look at Process Builder. We learned the difference between automation tools such as Flow, Process Builder, and Workflow Rule. We also discussed the various actions available in Process Builder. In this chapter, we will discuss the various ways to distribute Processes. We will also discuss how you debug your Process and forward a Flow or Process error email to other users or the internal IT team. The following topics will be covered in this chapter:

- Distributing or deploying Processes
- Debugging your Process
- Designating a recipient to receive Process error emails

Before going ahead, let me remind you that whenever you create a process, the system automatically creates a **Flow,** and a **Flow Trigger** to call the Flow. This happens behind the scenes, and the user doesn't need to interact with the shadow Flows. In this chapter, if you see the word Flow, do not get confused; we are talking about Process Builder, not Visual Workflow.

Distributing or deploying Processes

Once you are done with Process development, the next step is to deploy the Process. There are a couple of ways in which you can deploy or distribute it. They are as follows:

- Change Sets
- Packages
- Integrated development environment (such as `Force.com` IDE and MavnesMate)

Deploying using Change Sets

Change Sets allow you to deploy the Processes to a connected Salesforce organization such as your production environment or another Sandbox.

Let's look at a business scenario. Alice Atwood is working as a system administrator at Universal Containers. She has developed a process in a Sandbox (Full Sandbox) and is done with testing. She wants to migrate the newly created process to the production organization.

For the preceding business scenario, when both Salesforce organizations are connected like a Full Sandbox and Production, it's best practice to use Change Sets to deploy the components. In this chapter, we are going to focus on the alternative: using a package.

You can learn more about deploying Processes using Change Sets at `https://help.salesforce.com/HTViewHelpDoc?id=changesets.htm&language=en_US`.

Change Sets allow you to add active Processes to outbound Change Sets. Active Processes are available under the **Flow Definition** component type.

Hands on 1 – creating an unmanaged package

Packages give you the flexibility to deploy Processes in any Salesforce organization, which means that if you have developed a Process in a free developer organization and want to share it with your colleagues, they can install it in their Salesforce developer organization. In this case, the two Salesforce organizations are not connected.

To learn more about packages in Salesforce, visit `https://help.salesforce.com/HTViewHelpDoc?id=sharing_apps.htmlanguage=en_US`.

Let's look at a business scenario. Joe Thompson is working as a system administrator at Universal Containers. He has developed the `Auto create new Contract` process in `Chapter 1`, *Getting Started with Lightning Process Builder*, in his personal developer organization. Now he wants to install this process in his organization's Full Sandbox.

When the organizations are not connected, you can use a package to migrate the application, code, or any other changes. If a package (**Managed - Released**) contains Apex code, Flows, or Processes, then you can only install the packages in Developer Edition, Lightning Enterprise Edition, or higher, if the package doesn't pass the AppExchange security review. If a package that contains Apex code passes the security review, then you can install this type of application in any of the Salesforce editions, even in Lightning Professional Edition or below. To solve the preceding business requirement, we will create an unmanaged package. Perform the following steps to create a package:

1. In Lightning Experience, navigate to **Setup** (**Gear Icon**) | **Setup** | **PLATFORM TOOLS** | **Apps** | **Package Manager**, click on the **Packages** section, and then click on the **New** button.
2. It will redirect you to a new window, where you have to enter **Package Name**, **Language**, **Configure Custom Link**, **Notify on Apex Error**, and **Description**, as shown in the following screenshot:

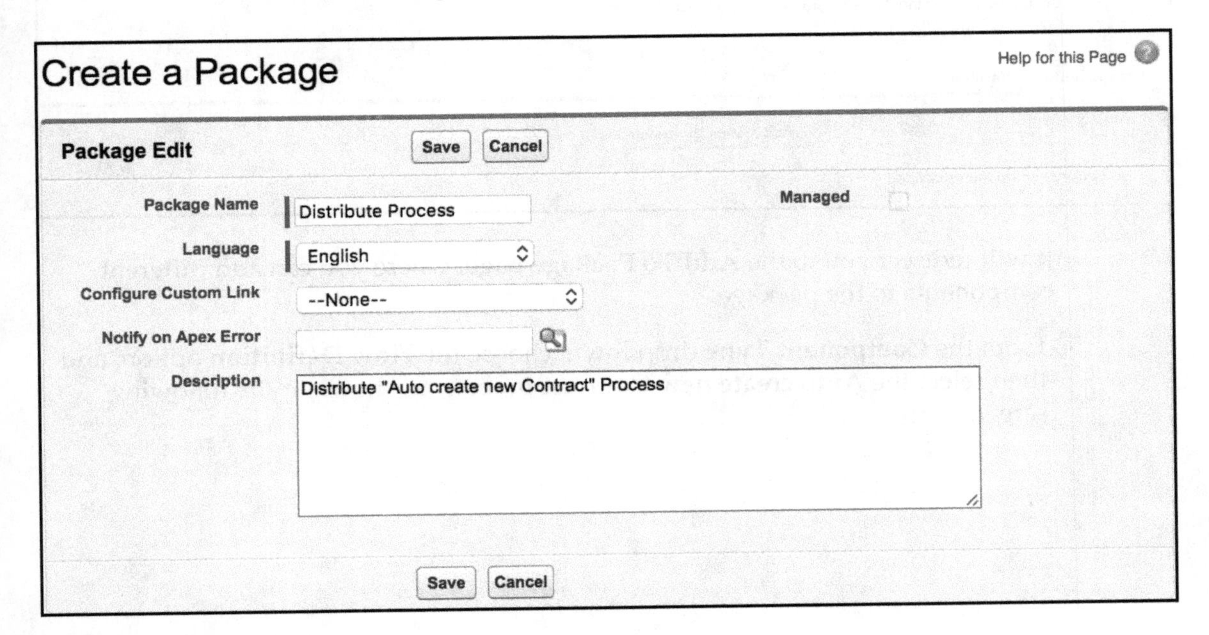

3. Once you are done, click on the **Save** button; it will redirect you to the **Package Detail** page.
4. The next task is to add components to the package. To do that, click on the **Components** tab, and then click on the **Add** button, as shown in the following screenshot:

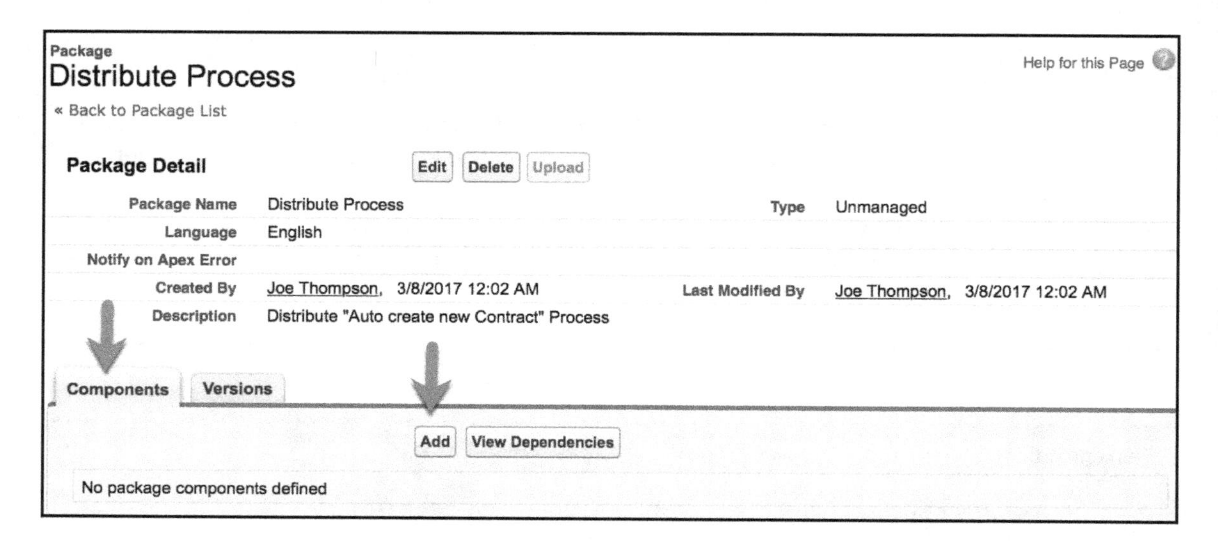

It will redirect you to the **Add To Package** page, where you can add different components to the package.

5. From the **Component Type** dropdown, choose the **Flow Definition** option, and then select the **Auto create new Contract** process, as shown in the following screenshot:

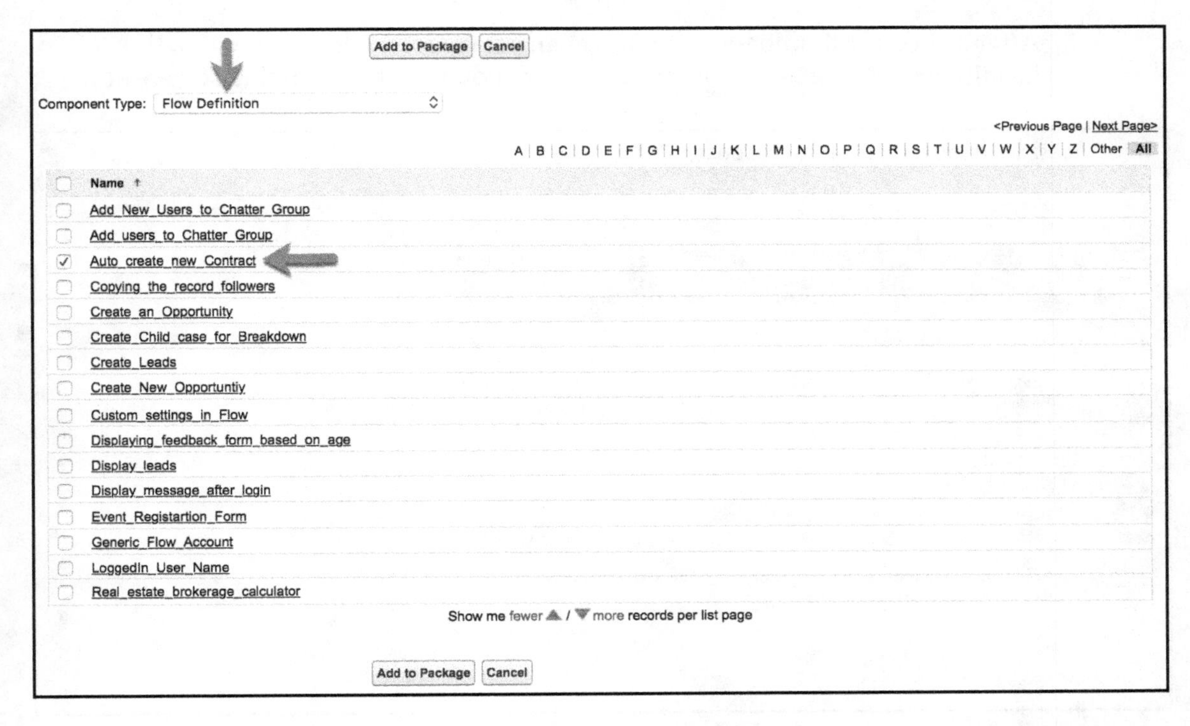

6. Once you are done, click on the **Add To Package** button. It will redirect you to the **Package Detail** page. Make sure that you have added all the dependent components that are used in Process Builder.
7. Once you are done, the next step is to upload the package. To do that, click on the **Upload** button; it will redirect you to the **Upload Package** page, where you have to enter the version name as `Summer2018`, `2.0` for the version number, and the requirements to install a package. Once you are done, click on the **Upload** button.

8. After a successful upload, you will get an email from Salesforce with a link to install the package in any organization, as shown in the following screenshot:

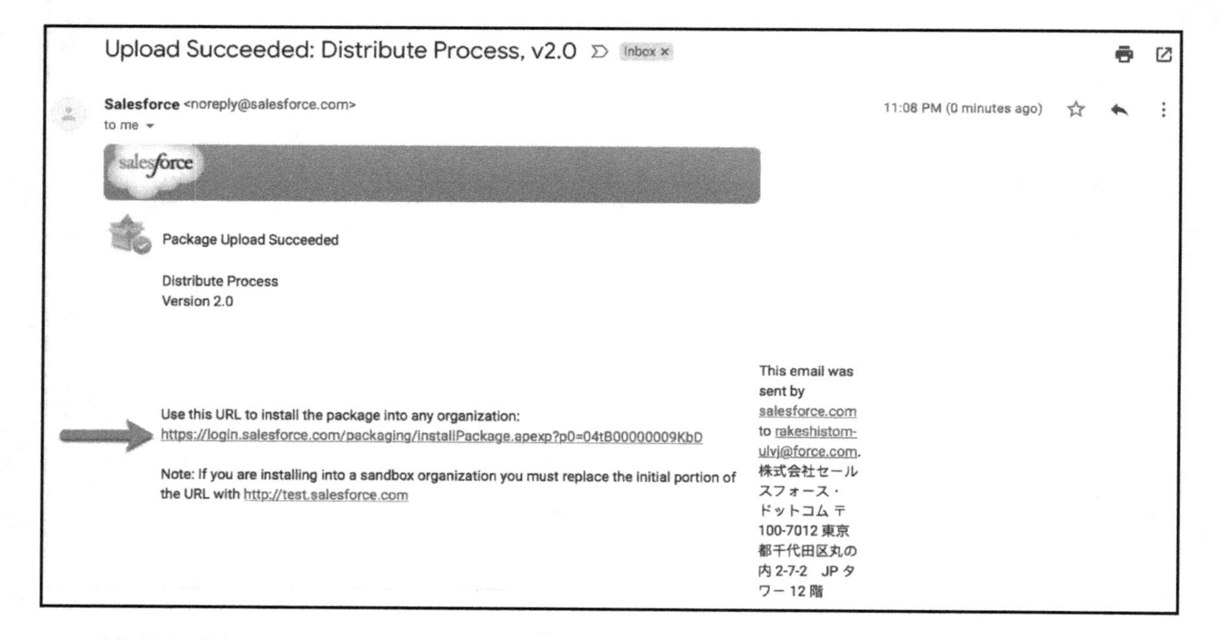

If you're planning to install this package in a developer organization, use the link as it is. To install this package in a Sandbox, replace `login` with `test`; it will then look like the following URL:

```
https://test.salesforce.com/packaging/installPackage.apexp?p0=04tB00000009KbD
```

We have created an unmanaged package to distribute the Processes. Using this URL, anyone will be able to install the package in their organization. You can make it secure by adding a password at the time of package creation.

Debugging your Process

While testing Processes, or following deployment, users may get an error. The next step is to find out how to debug a Process to understand why it's failing at runtime. Now we will discuss the various ways through which you can debug the Process.

Inbuilt error-displaying tool – on-screen debugging

While working, you may occasionally get an error generated by Process Builder, which may be too general, for example, **We're sorry but a serious error occurred. Please contact Salesforce Customer Support as soon as possible**. Salesforce displays friendlier and include the process name and Error ID. If the error occurs when creating or updating a record, the message also includes technical details for the user to assist you in troubleshooting. This is the first level of debugging, which allows you to find the root cause of the error.

When a Process fails, it will display a screen that includes the error message. This will appear as shown in the following screenshot:

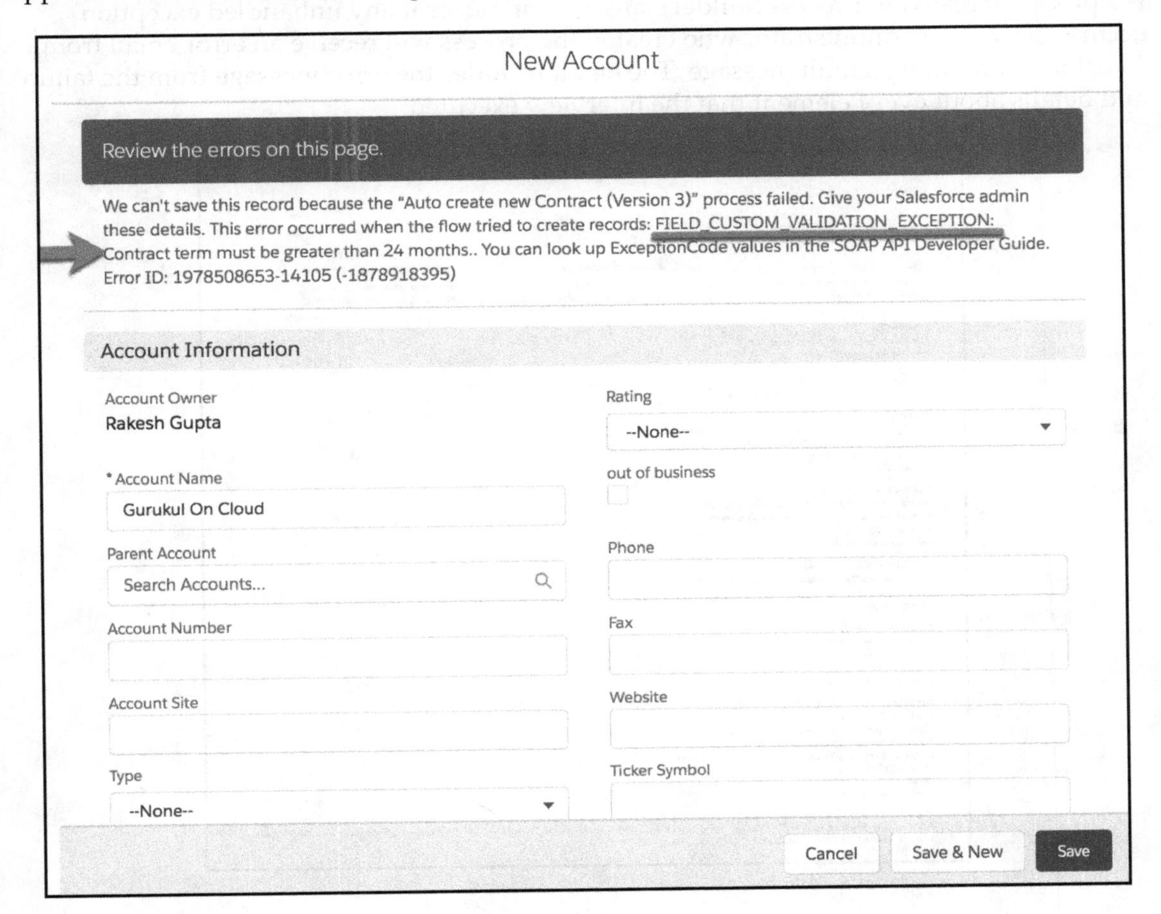

If you look at the screen carefully, you can easily understand why this happened. The error occurred because of a Validation Rule that doesn't allow anyone to create or update contracts with a term of less than 24 months. In our case, when a contract is automatically created by Process, the contract term is set at 12 months. There are two possible solutions:

- Modify the Process (**Auto create a Contract**) to update the **Create a Record** action and subsequently update the contract terms to greater than 24 months.
- Bypass the Process for a specific user or profile, which we will discuss in the next chapter.

Using process fault error emails

If a process (created in Process Builder) fails at runtime, or if any unhandled exception occurs, the system administrator who created the process will receive an error email from Salesforce containing a fault message. The email includes the error message from the failure and details about every element that the interview executed:

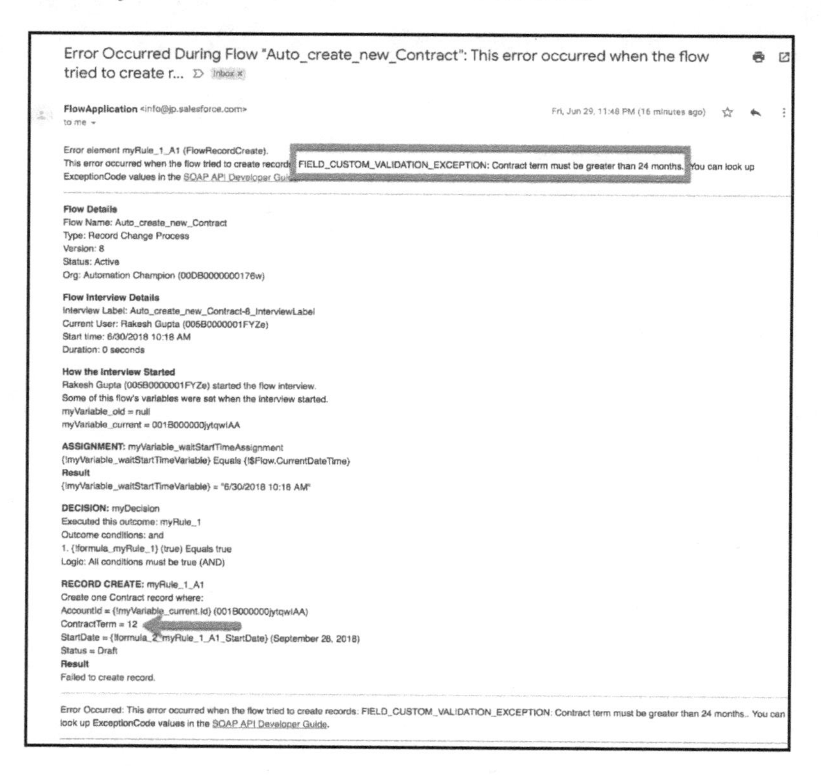

Error Occurred During Flow "Auto_create_new_Contract": This error occurred when the flow tried to create r... Inbox

FlowApplication <info@jp.salesforce.com>
to me
Fri, Jun 29, 11:48 PM (16 minutes ago)

Error element myRule_1_A1 (FlowRecordCreate).
This error occurred when the flow tried to create records: FIELD_CUSTOM_VALIDATION_EXCEPTION: Contract term must be greater than 24 months. You can look up ExceptionCode values in the SOAP API Developer Guide.

Flow Details
Flow Name: Auto_create_new_Contract
Type: Record Change Process
Version: 8
Status: Active
Org: Automation Champion (00DB0000000176w)

Flow Interview Details
Interview Label: Auto_create_new_Contract-8_InterviewLabel
Current User: Rakesh Gupta (005B0000001FYZe)
Start time: 6/30/2018 10:18 AM
Duration: 0 seconds

How the Interview Started
Rakesh Gupta (005B0000001FYZe) started the flow interview.
Some of this flow's variables were set when the interview started.
myVariable_old = null
myVariable_current = 001B000000jytqwIAA

ASSIGNMENT: myVariable_waitStartTimeAssignment
{!myVariable_waitStartTimeVariable} Equals {!$Flow.CurrentDateTime}
Result
{!myVariable_waitStartTimeVariable} = "6/30/2018 10:18 AM"

DECISION: myDecision
Executed this outcome: myRule_1
Outcome conditions: and
1. {!formula_myRule_1} (true) Equals true
Logic: All conditions must be true (AND)

RECORD CREATE: myRule_1_A1
Create one Contract record where:
AccountId = {!myVariable_current.Id} (001B000000jytqwIAA)
ContractTerm = 12
StartDate = {!formula_2_myRule_1_A1_StartDate} (September 28, 2018)
Status = Draft
Result
Failed to create record.

Error Occurred: This error occurred when the flow tried to create records: FIELD_CUSTOM_VALIDATION_EXCEPTION: Contract term must be greater than 24 months.. You can look up ExceptionCode values in the SOAP API Developer Guide.

The preceding screenshot displays a Process error generated at runtime because of the Process trying to create a Contract record with a **Contract term** of 12 months. If the system administrator who created the Flow or Process leaves the organization (that is, if the user gets deactivated in Salesforce), and if a Process fails at runtime or if any unhandled exception occurs, the inactive user will receive an error email from Salesforce containing a fault message. In this chapter, we will discuss how to overcome this challenge.

Using Debug Logs

The best way to troubleshoot runtime issues is to use **Debug Logs**. In the debug log, you will see the Flow action events in the workflow category of **Debug Logs**, which shows the Flow version and the values passed to it.

Before going ahead with the debug log, the one thing you have to do is set the correct log filters, the steps for which are as follows:

1. Navigate to **Setup** | **Monitor** | **Logs** | **Debug Logs**.
2. Click on **New,** available under the **Monitored Users** related list. Using a magnifying glass, add the users whose debug logs you would like to monitor and retain. In the current case, select your name.
3. Click on **Save**.
4. Navigate to the **Monitored Users** related list and click on the **Filters** link to change the debug logs filter. There are mainly three kinds of setting available:
    - **Log category**: The type of information logged.
    - **Log level**: For the Workflow and Visualforce page, set it as **FINER**. If you set the log level for the running user to be **FINER**, it will show your assignment actions and values.
    - **Events**: This is the combination of **Log category** and **Log level**.
5. Click on **Save**.

Let's take the preceding example. Go to the debug log and open it. You can easily identify the reason behind it. The best part of **Debug logs** is that you can see what is happening behind the scenes and you will get more information, such as the number of DML and SOQL queries used:

```
43.0 APEX_CODE,FINEST;APEX_PROFILING,FINEST;CALLOUT,FINEST;DB,FINEST;NBA,INFO;SYSTEM,FINE;VALIDATION,INFO;VISUALFORCE,FINER;WAVE,INFO;WORKFLOW,FINER
11:00:23.0 (454617)|USER_INFO|[EXTERNAL]|005B0000001FY2e|rakeshistom-ulvj@force.com|India Standard Time|GMT+05:30
11:00:23.0 (595448)|EXECUTION_STARTED
11:00:23.0 (605826)|CODE_UNIT_STARTED|[EXTERNAL]|Workflow:Account
11:00:23.0 (12437764)|WF_RULE_EVAL_BEGIN|Workflow
11:00:23.0 (12795945)|WF_CRITERIA_BEGIN|[Account: Gurukul On Cloud 001B000000jytyM]|RakeshGupta_Update_Contacts_Asst_Phone301B0000000ChJI|01QB00000008aDm|ON_ALL_CHANGES|0
11:00:23.0 (13058831)|WF_FORMULA|Formula:ENCODED:[treatNullAsNull]true|Values:
11:00:23.0 (13180873)|WF_CRITERIA_END|true
11:00:23.0 (13878197)|WF_CRITERIA_BEGIN|[Account: Gurukul On Cloud 001B000000jytyM]|RakeshGupta_Auto_create_new_Contract301B0000000ChJS|01QB0000000D9IA|ON_CREATE_ONLY|0
11:00:23.0 (14010647)|WF_FORMULA|Formula:ENCODED:[treatNullAsNull]true|Values:
11:00:23.0 (14027753)|WF_CRITERIA_END|true
11:00:23.0 (14098455)|WF_SPOOL_ACTION_BEGIN|Workflow
11:00:23.0 (14254691)|WF_ACTION| Flow Trigger: 2;
11:00:23.0 (14326844)|WF_RULE_EVAL_END
11:00:23.0 (15694156)|WF_FLOW_ACTION_BEGIN|09LB00000009sZ3
11:00:23.0 (15783842)|WF_FLOW_ACTION_DETAIL|09LB00000009sZ3|[Account: Gurukul On Cloud 001B000000jytyM]|Id=09LB00000009sZ3|CurrentRule:RakeshGupta_Auto_create_new_Contract301
11:00:23.19 (19403745)|FLOW_CREATE_INTERVIEW_BEGIN|00DB0000000176w|300B00000004HSn|301B0000000ChJS
11:00:23.19 (20139162)|FLOW_CREATE_INTERVIEW_END|462589952b6a6914a3711ee37cd1644e7d5973-48b5|Auto create new Contract (Version 3)
11:00:23.0 (20774329)|WF_FLOW_ACTION_DETAIL|Param Name: myVariable_current, Param Value: ENCODED:{![treatNullAsNull]{!ID:this}}, Evaluated Param Value: {Entity type: Account,
11:00:23.21 (21271055)|FLOW_START_INTERVIEWS_BEGIN|1
11:00:23.21 (21406686)|FLOW_START_INTERVIEW_BEGIN|462589952b6a6914a3711ee37cd1644e7d5973-48b5|Auto create new Contract (Version 3)
11:00:23.21 (21946244)|FLOW_START_INTERVIEW_LIMIT_USAGE|SOQL queries: 0 out of 100
11:00:23.21 (22101942)|FLOW_START_INTERVIEW_LIMIT_USAGE|SOQL query rows: 0 out of 50000
11:00:23.21 (22179857)|FLOW_START_INTERVIEW_LIMIT_USAGE|SOSL queries: 0 out of 20
11:00:23.21 (22267405)|FLOW_START_INTERVIEW_LIMIT_USAGE|DML statements: 0 out of 150
11:00:23.21 (22345355)|FLOW_START_INTERVIEW_LIMIT_USAGE|DML rows: 0 out of 10000
11:00:23.21 (22454058)|FLOW_START_INTERVIEW_LIMIT_USAGE|CPU time in ms: 0 out of 15000
11:00:23.21 (22579423)|FLOW_START_INTERVIEW_LIMIT_USAGE|Heap size in bytes: 0 out of 6000000
11:00:23.21 (22677836)|FLOW_START_INTERVIEW_LIMIT_USAGE|Callouts: 0 out of 100
11:00:23.21 (22744774)|FLOW_START_INTERVIEW_LIMIT_USAGE|Email invocations: 0 out of 10
11:00:23.21 (22865429)|FLOW_START_INTERVIEW_LIMIT_USAGE|Future calls: 0 out of 50
11:00:23.21 (22932877)|FLOW_START_INTERVIEW_LIMIT_USAGE|Jobs in queue: 0 out of 50
11:00:23.21 (23023623)|FLOW_START_INTERVIEW_LIMIT_USAGE|Push notifications: 0 out of 10
11:00:23.21 (23665456)|FLOW_VALUE_ASSIGNMENT|462589952b6a6914a3711ee37cd1644e7d5973-48b5|myVariable_old|
11:00:23.21 (24437911)|FLOW_VALUE_ASSIGNMENT|462589952b6a6914a3711ee37cd1644e7d5973-48b5|myVariable_current|{Id=001B000000jytyMIAQ, IsDeleted=false, MasterRecordId=null, Name
11:00:23.21 (24739252)|FLOW_ELEMENT_BEGIN|462589952b6a6914a3711ee37cd1644e7d5973-48b5|FlowAssignment|myVariable_waitStartTimeAssignment
11:00:23.21 (27132835)|FLOW_ASSIGNMENT_DETAIL|462589952b6a6914a3711ee37cd1644e7d5973-48b5|myVariable_waitStartTimeVariable|ASSIGN|6/30/2018 11:00 AM
11:00:23.21 (27178597)|FLOW_VALUE_ASSIGNMENT|462589952b6a6914a3711ee37cd1644e7d5973-48b5|myVariable_waitStartTimeVariable|2018-06-30T05:30:23Z
11:00:23.21 (27491641)|FLOW_ELEMENT_END|462589952b6a6914a3711ee37cd1644e7d5973-48b5|FlowAssignment|myVariable_waitStartTimeAssignment
11:00:23.21 (27523905)|FLOW_ELEMENT_BEGIN|462589952b6a6914a3711ee37cd1644e7d5973-48b5|FlowDecision|myDecision
11:00:23.21 (27775241)|FLOW_RULE_DETAIL|462589952b6a6914a3711ee37cd1644e7d5973-48b5|myRule_1|true
11:00:23.21 (27801165)|FLOW_VALUE_ASSIGNMENT|462589952b6a6914a3711ee37cd1644e7d5973-48b5|myRule_1|true
11:00:23.21 (29778520)|FLOW_ELEMENT_END|462589952b6a6914a3711ee37cd1644e7d5973-48b5|FlowDecision|myDecision
11:00:23.21 (29882837)|FLOW_ELEMENT_BEGIN|462589952b6a6914a3711ee37cd1644e7d5973-48b5|FlowRecordCreate|myRule_1_A1
11:00:23.21 (30076002)|FLOW_ELEMENT_DEFERRED|FlowRecordCreate|myRule_1_A1
11:00:23.21 (30132544)|FLOW_ELEMENT_END|462589952b6a6914a3711ee37cd1644e7d5973-48b5|FlowRecordCreate|myRule_1_A1
11:00:23.21 (30172648)|FLOW_START_INTERVIEW_END|462589952b6a6914a3711ee37cd1644e7d5973-48b5|Auto create new Contract (Version 3)
11:00:23.21 (30625860)|FLOW_BULK_ELEMENT_BEGIN|FlowRecordCreate|myRule_1_A1
11:00:23.21 (48267060)|LIMIT_USAGE|[EXTERNAL]|FIELDSETS_DESCRIBES|1|100
11:00:23.21 (48403423)|LIMIT_USAGE|[EXTERNAL]|FIELDSETS_DESCRIBES|2|100
11:00:23.21 (49496649)|FLOW_BULK_ELEMENT_DETAIL|FlowRecordCreate|myRule_1_A1|1
```

Through the debug log, you can easily track a bug or error in your Flow.

If a process fails at runtime, process creators receive an email from Salesforce with a fault message. If the interview failed at multiple elements, process creators receive multiple emails, and the final email includes an error message for each failure.

Setting up recipients to receive process error emails

Salesforce offers a handful of debugging tools; for instance, a user can use an error email sent by Salesforce, or leverage a standard debug log. Each option, however, has its pros and cons.

By default, when a Process fails, Salesforce sends a detailed error email to the Salesforce administrator who last modified the Process, as shown in the preceding screenshot.

Let's look at a business scenario. Alice Atwood is working as a system administrator at Universal Containers. She has created many Processes to streamline business processes. As she created all Flows and Processes, by default, she receives all error emails for Flows and Processes. Recently, she onboarded a new Salesforce Administrator, Sarika Gupta. Now, Alice is looking for a way to channel error emails to Sarika Gupta for all the Flows and Processes that she created.

Apex exception email recipients

Using Apex exception email recipients, you can configure who will receive emails when your Process, Flow, and Apex code encounters unhandled exceptions.

Follow these instructions to configure the Apex reception email recipients for the preceding business requirement:

1. To change the destination of Flow and Process error emails, navigate to **Setup (Gear Icon) | Setup | Process Automation | Process Automation Settings**.
2. Then, locate the **Send Process or Flow Error Email to** field, and update it to **Apex exception email recipients**.

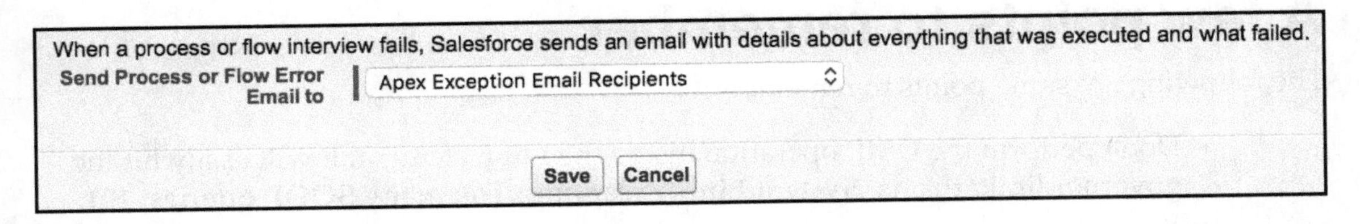

3. Once done, click on the **Save** button.
4. The next step is to specify users and email addresses as Apex exception email recipients. Navigate to **Setup (Gear Icon) | Setup | Email | Apex Exception Email**.

5. Click on the **Add Salesforce User** button to add a Salesforce user:

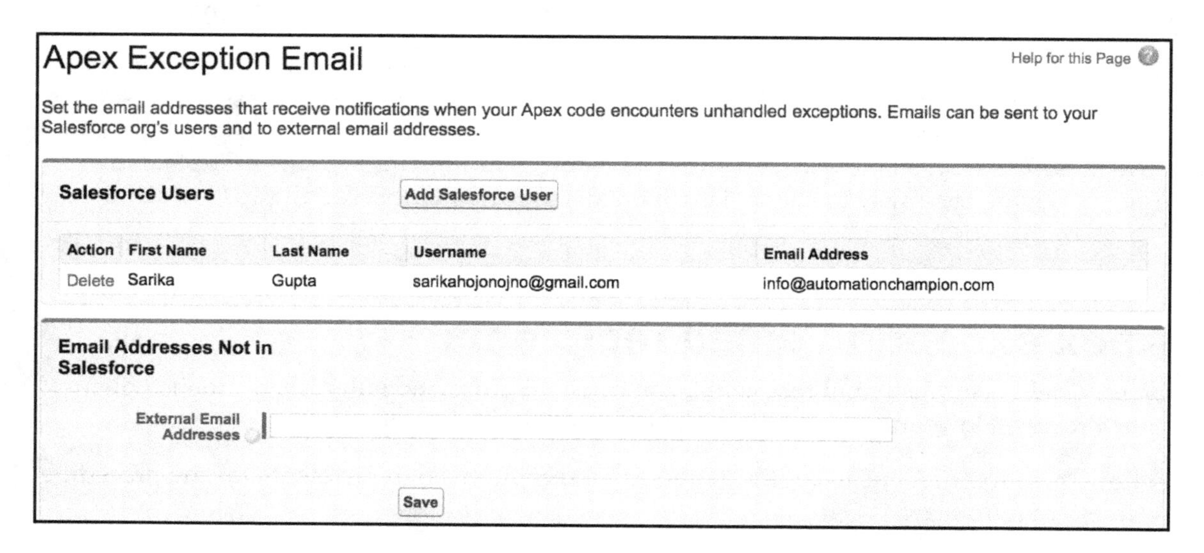

5. It is also possible to include an email address that doesn't belong to any user. For example, you may want to forward these emails to your IT team.

A few points to remember

The following are some points to remember:

- Don't perform the DML operation inside the **Loop** element. It will easily hit the governor limit, that is, **System.LimitException: Too many SOQL queries: 101**.
- You can only launch **Autolaunched Flow** from Process Builder.
- A Flow runs in user mode and Process Builder runs in system mode. Let's take an example—if you are trying to update an Opportunity, the next step would be as follows:
  - **If you use Process Builder**: If the running user doesn't have access to the `next step` field, Process Builder will be able to update it.
  - **If you use Flow**: If the running user doesn't have access to the `next step` field, they will get an error.

- The DML operation on a setup object is not permitted at the same time as when you update a non-setup object (or vice versa). If you want to do that, then use a time-dependent action.
- The total number of records Flow can retrieve using SOQL queries is 50,000, and only 10,000 will be processed as DML operations.
- You can have up to 50 versions of a Process, but only one version of a Process can be active.

Summary

In this chapter, we went through various concepts related to automating your business processes. We started the chapter with the concept of deploying or distributing the Flows or Processes. Then, we moved on to discuss the ways in which you can debug a Process. We also discussed a way to set up recipients for process error emails. In the next chapter, we will discuss concepts related to how you can create efficient and performance-optimized processes. We will also cover concepts such as executing multiple process criteria.

3
Building Efficient and Performance-Optimized Processes

In the previous chapter, we discussed how to deploy a Process or distribute it. We learned how to debug Process Builder. We also discussed how to forward a Flow or Process error email to other users or an IT team.

In this chapter, we will discuss some advanced concepts of Process Builder, such as how to apply a filter when updating related records, how to execute multiple groups of actions, and how to call Apex from Process Builder. We will also discuss how you can use one process to implement multiple business requirements.

The following topics will be covered in this chapter:

- Using Workbench to get the complete details of a process
- Using custom labels in Process Builder
- Calling an Apex class from Process Builder
- Bypass processes using custom permissions
- Defining additional conditions when updating records
- Scheduling multiple groups of actions
- Executing multiple criteria of a process
- Creating reusable processes by using an invocable process
- Using custom metadata types in a Flow

An overview of process management

Now that you have an understanding of Process Builder, you know that it is a tool that allows you to streamline business processes without writing code. You can also use it to view your business process visually. The process management page allows you to see all of the processes in the current Salesforce organization. The process management page provides the following options:

- Creating a new process.
- Editing a process.
- Deleting an inactive process, or its version.
- Checking the statuses of processes that have been created for the current organization.
- Sorting the processes by name, description, object, last modified date, or status.

Your process management page should look like the following screenshot:

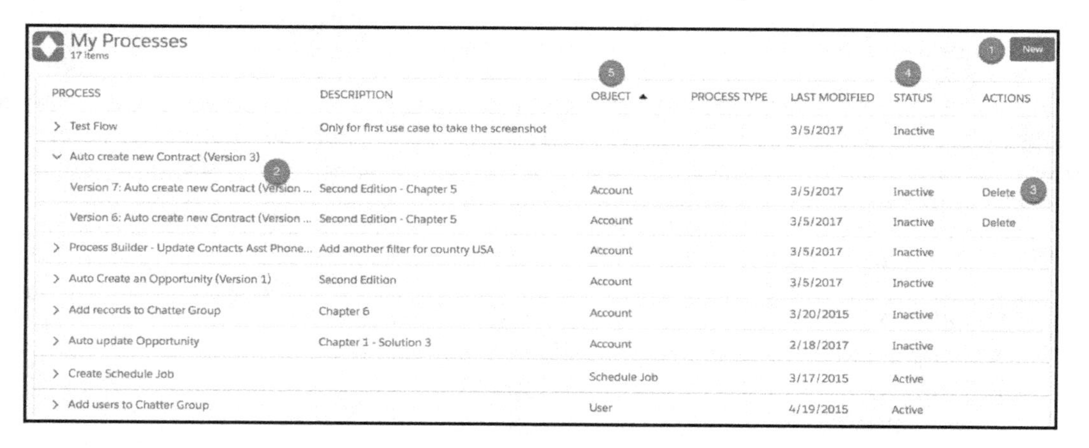

The process management page displays the fields: **PROCESS**, **DESCRIPTION**, **OBJECT**, **PROCESS TYPE**, **LAST MODIFIED**, **STATUS**, and **ACTIONS**. As of the Spring 2017 release, Process Builder does not display Created By, Created Date, or Last Modified By on the process management page.

Let's look at a business scenario. Suppose that Alice Atwood is working as a system administrator at Universal Containers. In `Chapter 1`, *Getting Started with Lightning Process Builder*, she developed a process (that is, `Update Contacts Asst Phone`) to fulfill the requirement that once an account gets activated, the related contact Asst. Phone field must be synced with the account's Phone.

She has received a few messages from business users, saying that the process stops when she tries to sync the contact Asst. Phone field with the account Phone field. While looking at the process version, she sees that someone has deactivated the process. Now, she wants to identify who deactivated the process.

Here are the methods that can be used to track a user that activated, deactivated, created, or deleted a Flow or Process:

- View the Setup Audit Trail
- Metadata component (Flow and FlowDefinition)

Using Audit Trail to track setup changes in a Process or a Flow

Setup Audit Trail tracks the recent configuration changes that you and other system administrators have made to your Salesforce organization. The audit history is especially useful in organizations with multiple system administrators. You can use the Audit Trail to track a user that activated, deactivated, created, or deleted a Flow or Process:

View Setup Audit Trail

Date	User	Action	Section	Delegate User
3/6/2017 6:32:36 PM PST	sarikahojonojno@gmail.com	Deleted Flow Trigger RakeshGupta_Update_Contacts_Asst_Phone301B0000000CfPs for Object: Account	Workflow Rule	
3/6/2017 6:32:36 PM PST	sarikahojonojno@gmail.com	Deleted workflow rule RakeshGupta_Update_Contacts_Asst_Phone301B0000000CfPs for Object: Account	Workflow Rule	
3/6/2017 6:32:36 PM PST	sarikahojonojno@gmail.com	Deactivated workflow rule RakeshGupta_Update_Contacts_Asst_Phone301B0000000CfPs for Object: Account	Workflow Rule	
3/6/2017 6:32:36 PM PST	sarikahojonojno@gmail.com	Deactivated flow version #4 "Process Builder - Update Contacts Asst Phone (Version 2)" for flow with Unique Name "Update_Contacts_Asst_Phone"	Flows	
3/6/2017 6:01:54 PM PST	alice.atwood@book.com	Created Flow Trigger RakeshGupta_Update_Contacts_Asst_Phone301B0000000CfPs for Object: Account	Workflow Rule	
3/6/2017 6:01:54 PM PST	alice.atwood@book.com	Created workflow rule RakeshGupta_Update_Contacts_Asst_Phone301B0000000CfPs for Object: Account	Workflow Rule	
3/6/2017 6:01:54 PM PST	alice.atwood@book.com	Activated flow version #4 "Process Builder - Update Contacts Asst Phone (Version 2)" for flow with Unique Name "Update_Contacts_Asst_Phone"	Flows	

In `Chapter 1`, *Getting Started with Lightning Process Builder*, we discussed that Process Builder is nothing but a combination of **Flow** (type: **Workflow**) and **Flow Trigger** (deprecated). Process Builder generates a Flow and Flow Trigger for all processes. This happens behind the scenes, and the user doesn't need to interact with these shadow Flows. The Audit Trail only stores the data for the past 180 days. So, it is helpful to track recent setup changes in Flows or Processes.

Go to `https://help.salesforce.com/articleView?id=admin_monitorsetup.htmlanguage=en_US` to learn more about Audit Trail.

Hands on 1 – using Workbench to get the details of a process

If you want to know details about the version of a process, or if you want to know about a process that someone created a few months or a year back, then the metadata component is the right option for you.

Let's look at a business scenario. Suppose that Alice Atwood is working as a system administrator at Universal Containers. In `Chapter 1`, *Getting started with Lightning with Process Builder*, she developed a process, **Update Contacts Asst Phone.** Now, she wants to identify the user that created version 1 of this process.

Before going on, let's look at the metadata components that store Flow and its versions in Salesforce, which are as follows:

- **Flow**: This represents the metadata that is associated with a Flow.
- **FlowDefinition**: This represents the Flow definition's description and the active Flow version number.

The following screenshot displays the Flow and Flow Definition:

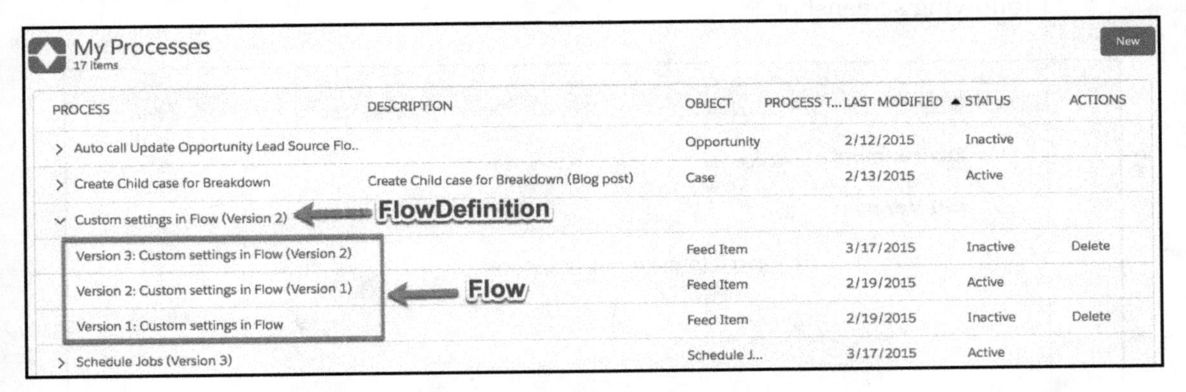

To view the metadata components, you can use either the `Force.com` IDE, the `Force.com` Migration Tool, or any third-party tool that supports viewing the metadata components of your Salesforce organization. **Workbench** is a powerful, web-based tool that's designed for administrators and developers to interact with their Salesforce organizations via the `Force.com` APIs. It also offers support for the Bulk API, Rest API, Streaming API, Metadata API, and Apex APIs, which let users describe things, query the objects, and manipulate the data in a Salesforce organization, directly in their web browser.

Follow these instructions to achieve the preceding tasks, using Workbench:

1. Log in to Workbench by going to `https://workbench.developerforce.com/login.php`.
2. For **Environment**, select **Production** (if you are using developer or live org) or **Sandbox** (if you are using sandbox). For **API Version**, select **37.0**, or the highest number. Make sure that you have selected the **I agree to the terms of service** checkbox.

3. Once you're done, click on the **Login with Salesforce** button, as shown in the following screenshot:

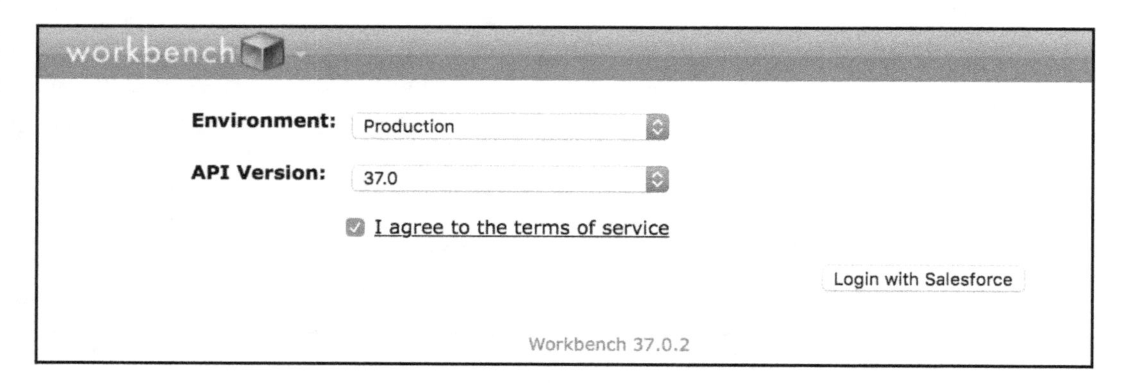

4. After successfully logging in, under the **Jump to** drop-down menu, select **Metadata Types & Components**, and then click on the **Select** button, as shown in the following screenshot:

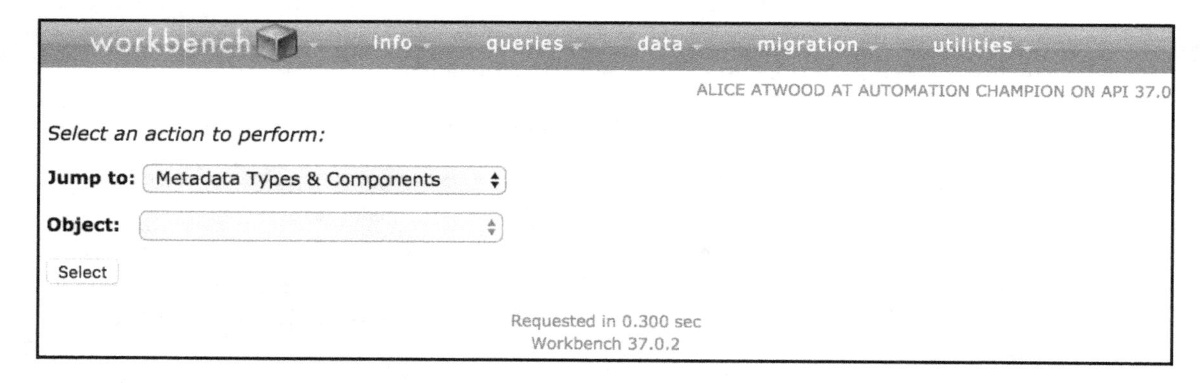

5. The next step is to choose the **Metadata Types & Components.** In this case, select **Flow**, and then click on **Components** to see the versions of all Flows and Processes, as shown in the following screenshot:

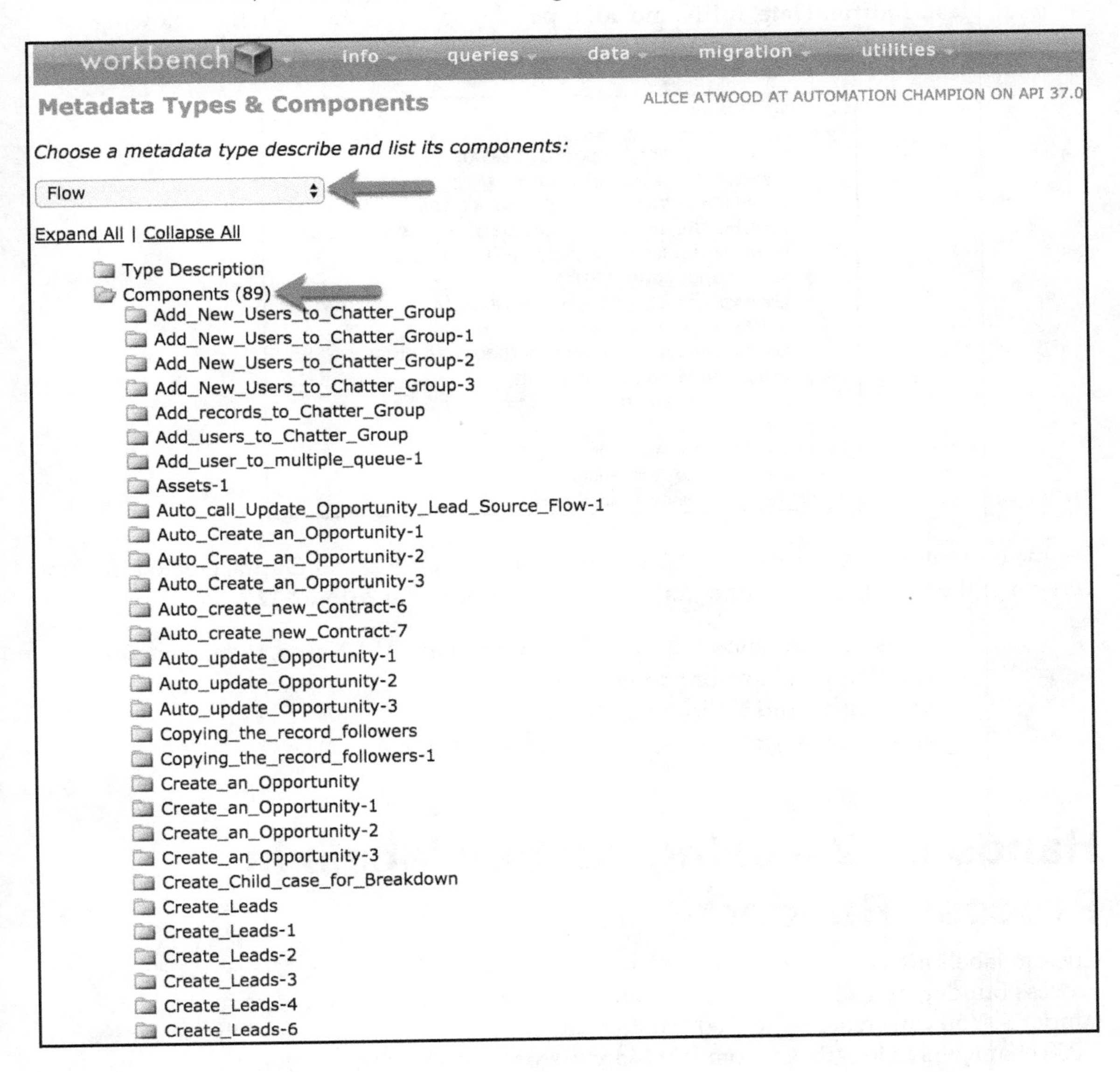

6. Now, identify the `Update_Contacts_Asset_Phone-1` Flow (the behind the scenes process is nothing but a combination of Flow and Flow Trigger). It will display the **createdByName**, **createdDate**, **lastModifiedByName**, **lastModifiedDate**, **fullName**, **id**, **type**, and so on, as shown in the following screenshot:

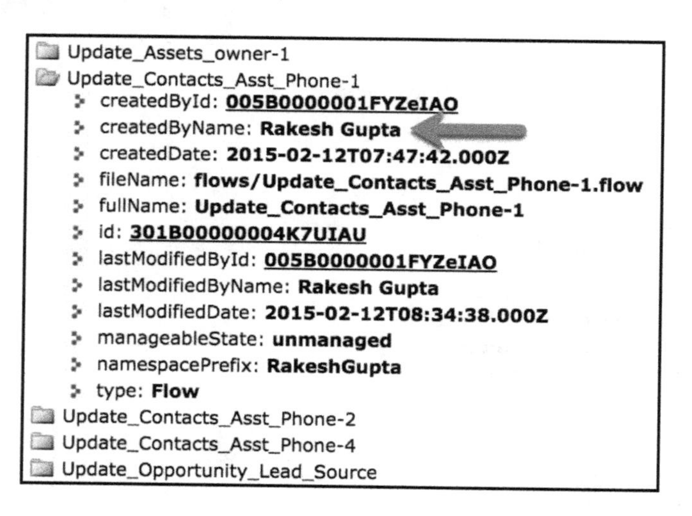

For the current business use case, the system administrator, **Rakesh Gupta,** created the first version of the `Update_Contacts_Asst_Phone-1` process on **2015-02-12**.

Salesforce appends `1`, `2`, `3`, and so on with the **API Name,** to manage the versions of a Flow or Process. For the preceding business scenario, the versions would be `Update_Contacts_Asset_Phone-1` (Version 1) and `Update_Contacts_Asset_Phone-2` (Version 2).

Hands on 2 – using custom labels in Process Builder

Custom labels are custom text values that can be accessed from Apex, Visual Workflow, Process Builder, and so on. The values can be translated into any language that Salesforce supports. You can create up to 5,000 custom labels in an organization, and they can be up to 1,000 characters in length. Custom labels are not only used for translation; they can also be used to store the username, password, and endpoint URL, in the case of invoking API calls for a third-party system.

For example, if you are integrating two systems, Salesforce and SAP, to sync the account information, in order to start the API calls, you will have to pass SAP integration user credentials and an endpoint URL. You will have three options to store these values:

- Hardcoding the credentials and endpoint URL in an Apex class
- Using multiple custom labels to save the username, password, and endpoint URL
- Using a custom metadata type to store this information

The benefit of using custom labels, compared to hardcoding the username and password in Apex, is that if the password changes in the future, you won't have to update the Apex class; instead, you can update the custom labels. This can be easily done in the production org itself. On the other hand, to update the Apex class, you have to use the Sandbox, and then migrate the changes into your production org. Using this approach, you can save time and hassle. You can also take the approach of using a custom metadata type, which is totally fine; but remember that Salesforce has put a limitation on everything. So, before starting the implementation, you should carefully design your complete process. I would suggest that you use custom labels if you want to save the credentials and the endpoint URL, as you wouldn't want to save this information in the cache.

Let's look at a business scenario. Suppose that Alice Atwood is working as a system administrator at Universal Containers. She has received a requirement from the management to auto-add an activated **Campaign** to the **Sales Executive** Chatter group.

Creating a custom label

We will now create a custom label to store the Sales Executive Chatter group ID. Perform the following steps to create a custom label to store the Chatter group ID:

1. First of all, navigate to the Chatter group **Sales Executive,** and copy the group ID from the URL. It will look like this: `0F9B0000000CyPe`.
2. Navigate to **Setup** (gear icon) | **Setup** | **PLATFORM TOOLS** | **User Interface** | **Custom Labels**, and click on the **New Custom Label** button; it will redirect you to a new window, where you will have to enter following details:
    - **Short Description**: Enter an easily identifiable term to recognize this custom label. In this case, use **Sales Executive Id**.
    - **Name**: This will be auto-populated, based on the **Short Description**.

- **Categories**: Enter the text to categorize the label. You can use this field in the filter criteria when creating custom label list views.
- **Value**: Enter text, up to 1,000 characters. In this case, enter the Sales Executive Chatter group ID, `0F9B0000000CyPe`.

This will look like the following screenshot:

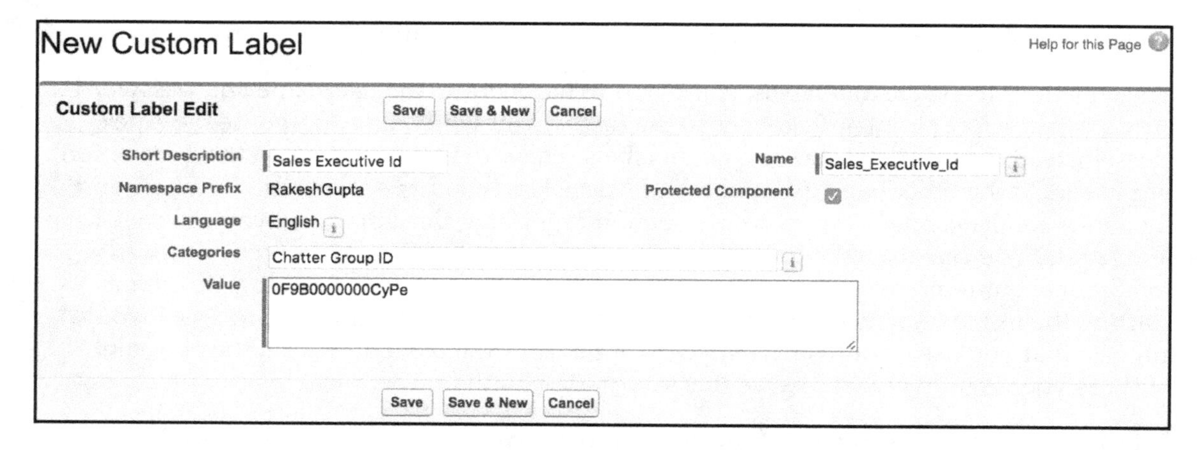

3. Once you are done, click on the **Save** button.

Hands on 3 – using a quick action to add a record to a Chatter group

Quick actions can be object-specific or global actions. To use a quick action in Process Builder, the action must exist in your organization. If your organization is using quick actions that allow your users to create and update records easily, you can also use these actions in Process Builder. When you use these quick actions in Process Builder, you are only allowed to set values for fields that are a part of the action's layout. Now, we will create a process to fulfill the preceding business scenario and use our custom label:

1. In **Lightning Experience**, navigate to **Setup** (gear icon) | **Setup** | **PLATFORM TOOLS** | **Feature Settings** | **Chatter** | **Chatter Settings,** and make sure that the **Allow Records in Groups** checkbox is selected.

2. To create a process, navigate to **Setup (Gear Icon)** | **Setup** | **PLATFORM TOOLS** | **Process Automation** | **Process Builder**, click on the **New** button, and enter the following details:
    - **Name**: Enter the name of the process - `Add campaign to Sales Executive group.`
    - **API Name**: This will be auto-populated, based on the name.
    - **Description**: Write some meaningful text, so that other developers or administrators can easily understand why this process was created.
    - **This process starts when**: Configure the process to start when a record is created or edited. In this case, select **A record changes.**
3. Once you are done, click on the **Save** button.
4. After **Define Process Properties,** the next task is to select the object upon which you want to create a process and define the **evaluation criteria**. For this, click on the **Add Object** node. This will open an additional window on the right-hand side of the process canvas screen, where you will have to enter the following details:
    - **Object**: Start typing, and then select the **Campaign** object.
    - **Start the process**: For **Start the process**, select **when a record is created or edited**. This means that the process will fire whenever a record gets created or edited.
    - **Recursion - Allow process to evaluate a record multiple times in a single transaction?**: Select this checkbox only when you want a process to evaluate the same record up to five times in a single transaction. In this case, leave this box unchecked.

The additional window will look as follows:

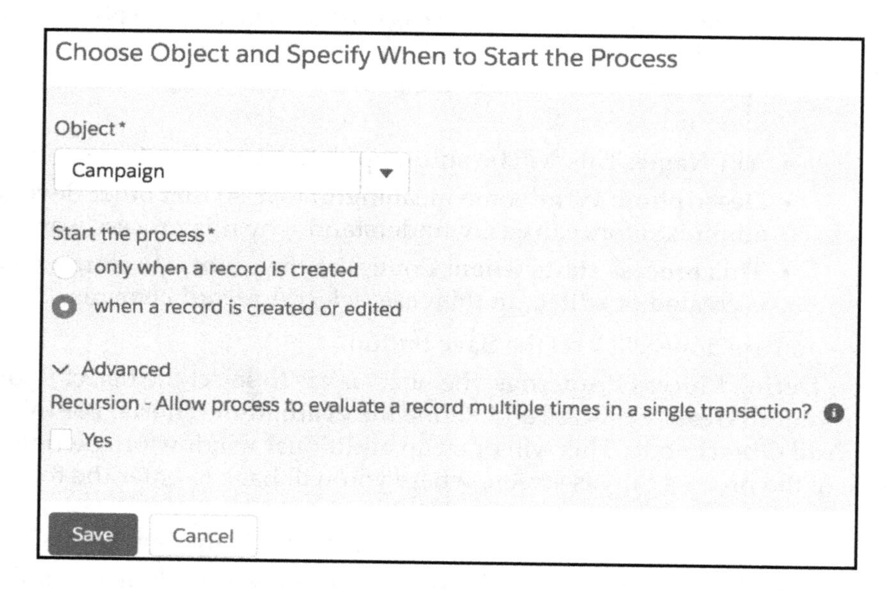

5. Once you are done, click on the **Save** button.
6. After defining the evaluation criteria, the next step is to add the **process criteria**. Once the process criteria are true, only then will the process execute the associated actions. To define the process criteria, click on the **Add Criteria** node. It will open an additional window on the right-hand side of the process canvas screen, where you will have to enter the following details:
    - **Criteria Name**: Enter a name for the criteria node. Enter **Only for active campaigns** as the criteria name, in this case.
    - **Criteria for Executing Actions**: Select the type of criteria that you want to define. You can select either **Formula evaluates to true,** or **Conditions are met** (a filter to define the process criteria), or **No criteria-just execute the actions!**. In this case, select **Conditions are met.**
    - **Set Conditions**: This field lets you specify which combination of the filter conditions must be true for the process to execute the associated actions. Set `[Campaign].IsActive` to `True`.
    - **Conditions**: In the **Conditions** section, select **All of the conditions are met (AND)**. This field lets you specify which combination of the filter conditions must be true for the process to execute the associated actions.

- Under **Advanced**, select **Yes** to execute the actions only when the specified changes are made.

The preceding actions will look as follows:

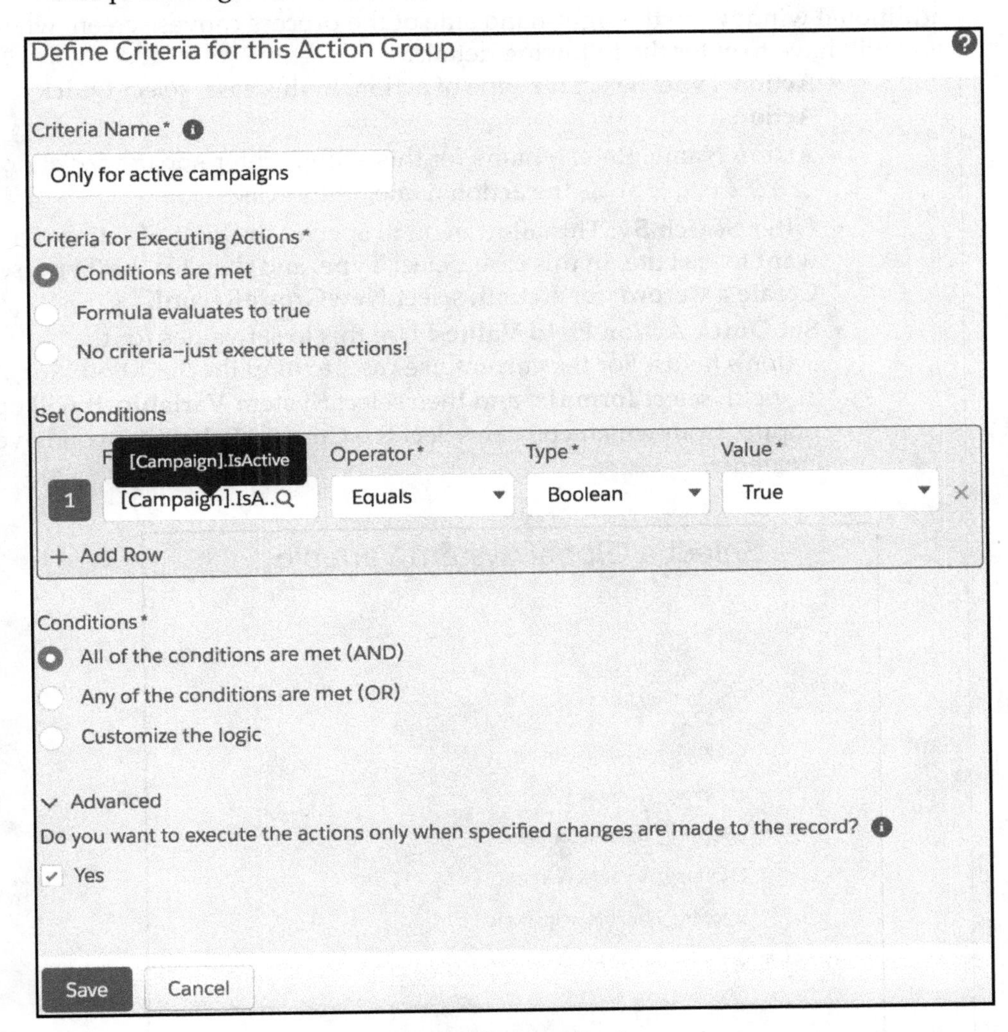

7. Once you are done, click on the **Save** button.

8. Once you are done with the **process criteria** node, the next step is to add an immediate action to add a campaign to the Sales Executive Chatter group. For this, we will use the **Quick Actions** action, available in Process Builder. Click on **Add Action,** available under **IMMEDIATE ACTIONS**; it will open an additional window on the right-hand side of the process canvas screen, where you will have to enter the following details:
    - **Action Type**: Select the type of action; in this case, select **Quick Actions**.
    - **Action Name**: Enter a name for this action. Enter `Add record to Chatter group` as the action name, in this case.
    - **Filter Search By**: This allows you to specify the kind of action that you want to execute. In this case, select **Type,** and then for the **Type**, select **Create a Record**; for **Action**, select **NewGroupRecord.**
    - **Set Quick Action Field Values**: Use this to set values for the action's fields. For the current use case, to map the `Related Record ID` field, select **formula**, and then select **System Variable.** It will open a popup, from which you can select the **Custom Label** that you have created:

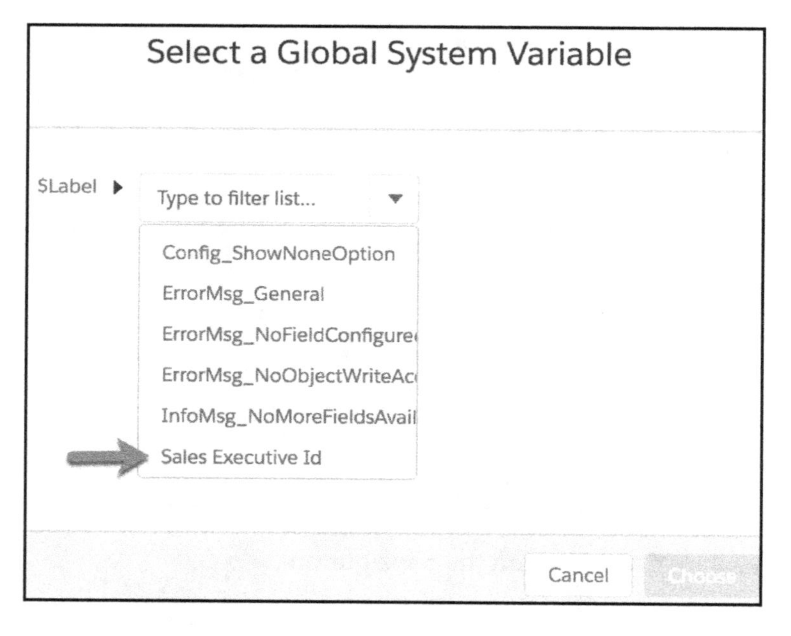

Likewise, map the `Record ID` field to `[Campaign].Id`.

The preceding steps will look as follows:

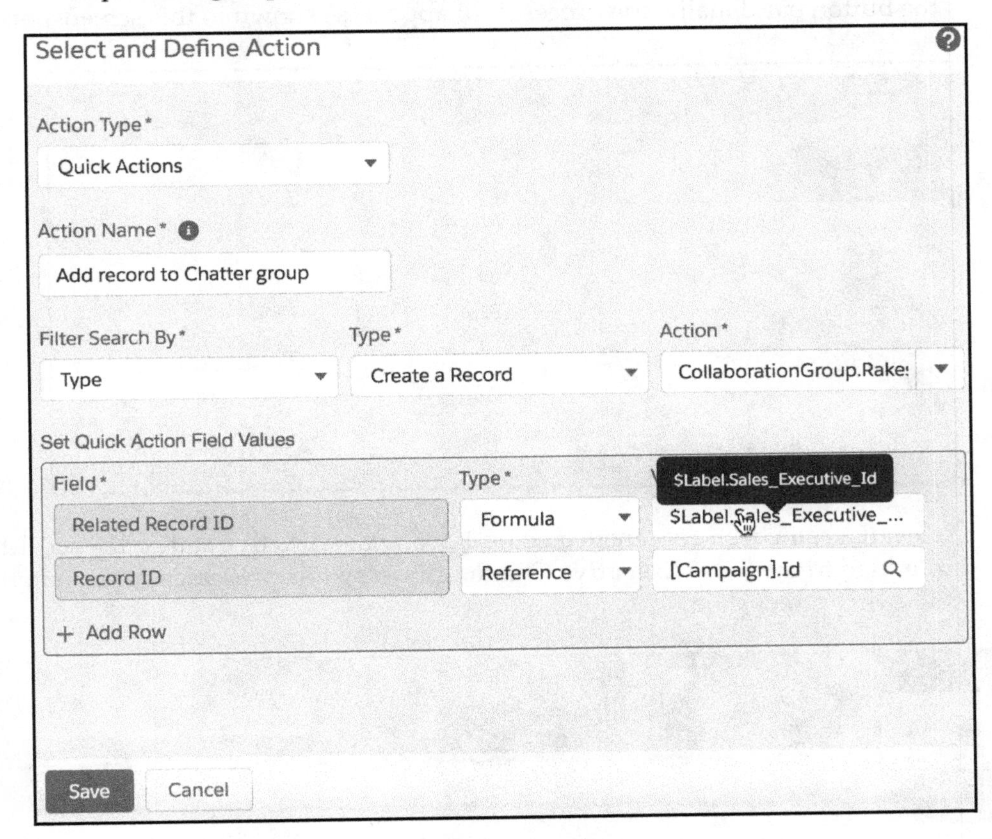

9. Once you are done, click on the **Save** button.

10. The final step is to activate the process. To do so, click on the **Activate** button, on the button bar. Finally, the process will appear, as shown in this screenshot:

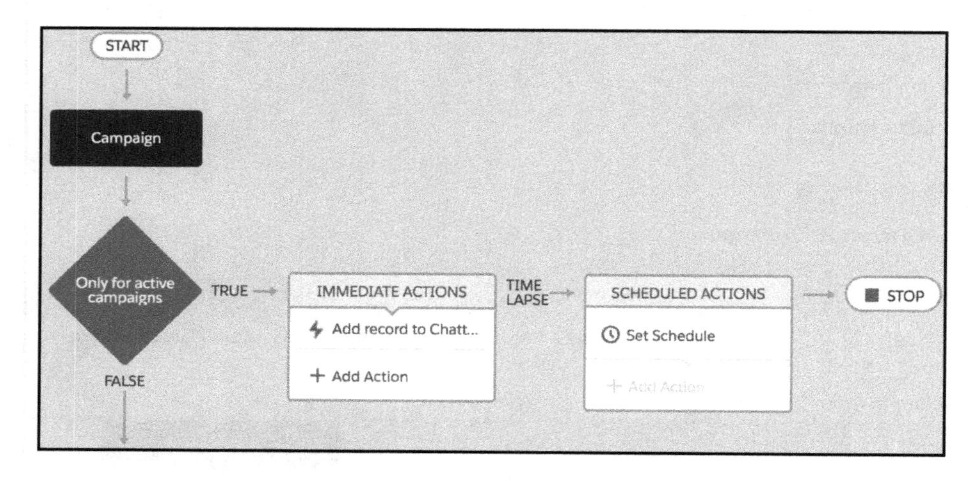

Now, if a campaign gets created or updated with an **Active** status, the record will be added to the **Sales Executive** Chatter group by this process, as shown in the following screenshot:

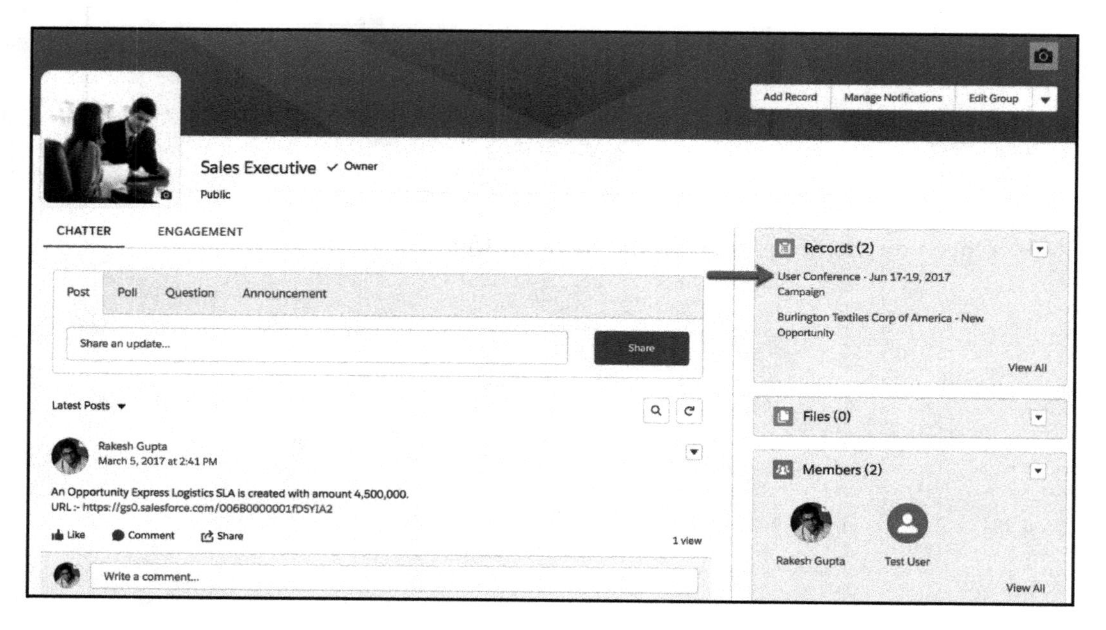

Likewise, you can create another process, to add new users to a Chatter group.

Hands on 4 – calling an Apex class from Process Builder

Process Builder allows you to call an Apex class that includes methods annotated with `@InvocableMethod`. By calling an Apex class from Process Builder, you can add customized functionalities, such as auto-converting leads, deleting records, or running lead assignment rules. You can also pass a required value into Apex class variables.

When no other process actions can get your job done, by calling an Apex method, you can add customized functionality for your users. You can only call an Apex class from Process Builder or Visual Workflow that have the `@InvocableMethod` annotation. This means that it is possible to extend the Process Builder functionality by writing an Apex class that executes your business logic, and then invoking the Apex from your process. If the class contains one or more invocable variables, then you have to manually enter values or reference field values from a related record.

Let's look at a business scenario. Suppose that Alice Atwood is working as a system administrator at Universal Containers. She has received a requirement from the management to auto-delete open cases, if the **out of business** checkbox is checked on the account record.

The following is the approach that we are going to use in order to solve the preceding business requirement:

1. First of all, create a custom **out of business** checkbox field on the **account** object, and make sure that you set the field-level security for the respective profiles.
2. The next step is to write an Apex class that will delete open cases for accounts whose IDs we are going to pass through the process. To create an Apex class, navigate to **Setup** (gear icon) | **Setup** | **PLATFORM TOOLS** | **Custom Code** | **Apex Classes,** and then click on the **New** button. The following is sample code; `AccountIds` is the record ID of accounts where out of business is updated as **true**:

```
public class DeleteOpenCases
{
    @InvocableMethod
    public static void CaseDelete(List<Id> AccountIds)
    {
        List<Case> Cases =[select id from case
                           where Account.id in :AccountIds
                           and Status != 'Closed'];
        delete Cases;
```

```
        }
    }
```

3. The next step is to create a process to call the Apex class that you have just created, only when the **out of business** checkbox is checked on the account record. To create a process, navigate to **Setup** (gear icon) | **Setup** | **PLATFORM TOOLS** | **Process Automation** | **Process Builder**, click on the **New** button, and enter the following details:
    - **Name**: Enter the name of the process - `Delete open cases - Out of business accounts.`
    - **API Name**: This will be auto-populated, based on the name.
    - **Description**: Write some meaningful text, so that other developers or administrators can easily understand why this process has been created.
    - **This process starts when**: Configure the process to start when a record is created or edited. In this case, select **A record changes.**
4. Once you are done, click on the **Save** button.
5. After **Define Process Properties**, the next task is to select the object upon which you want to create a process and define the **Evaluation Criteria**. For this, click on the **Add Object** node. It will open an additional window on the right-hand side of the process canvas screen, where you will have to enter the following details:
    - **Object**: Start typing, and then select the **Account** object.
    - **Start the process**: For **Start the process**, select **when a record is created or edited**. This means that the process will fire whenever a record gets created or edited.
    - **Recursion - Allow process to evaluate a record multiple times in a single transaction?**: Select this checkbox only when you want the process to evaluate the same record up to five times in a single transaction. In this case, leave this box unchecked.

The preceding steps will look like the following screenshot:

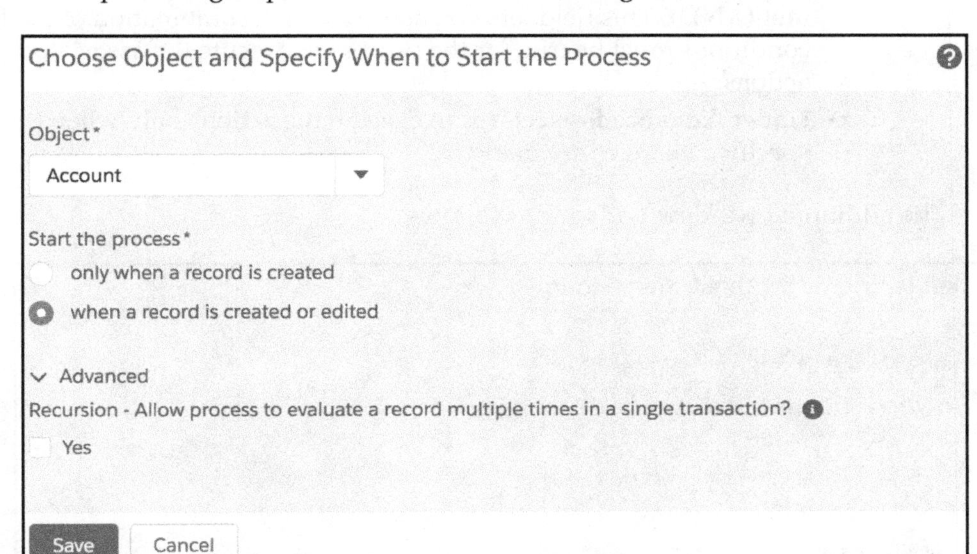

6. Once you are done, click on the **Save** button.
7. After defining the **Evaluation Criteria**, the next step is to add the **Process Criteria**. Once the process criteria are true, the process will execute the associated actions. To define the process criteria, click on the **Add Criteria** node. It will open an additional window on the right-hand side of the process canvas screen, where you will have to enter the following details:
    - **Criteria Name**: Enter a name for the criteria node. Enter `Only for out of business accounts` as the criteria name, in this case.
    - **Criteria for Executing Actions**: Select the type of criteria that you want to define. You can select either **Formula evaluates to true,** or **Conditions are met** (a filter to define the process criteria), or **No criteria-just execute the actions!**. In this case, select **Conditions are met.**
    - **Set Conditions**: This field lets you specify which combination of the filter conditions must be true for the process to execute the associated actions. In this case, set `[Account].RakeshGupta__out_of_business__c` to `True`.

- **Conditions**: In the **Conditions** section, select **All of the conditions are met (AND)**. This field lets you specify which combination of the filter conditions must be true for the process to execute the associated actions.
- Under **Advanced**, select **Yes** to execute the actions only when the specified changes are made.

The additional window will look as follows:

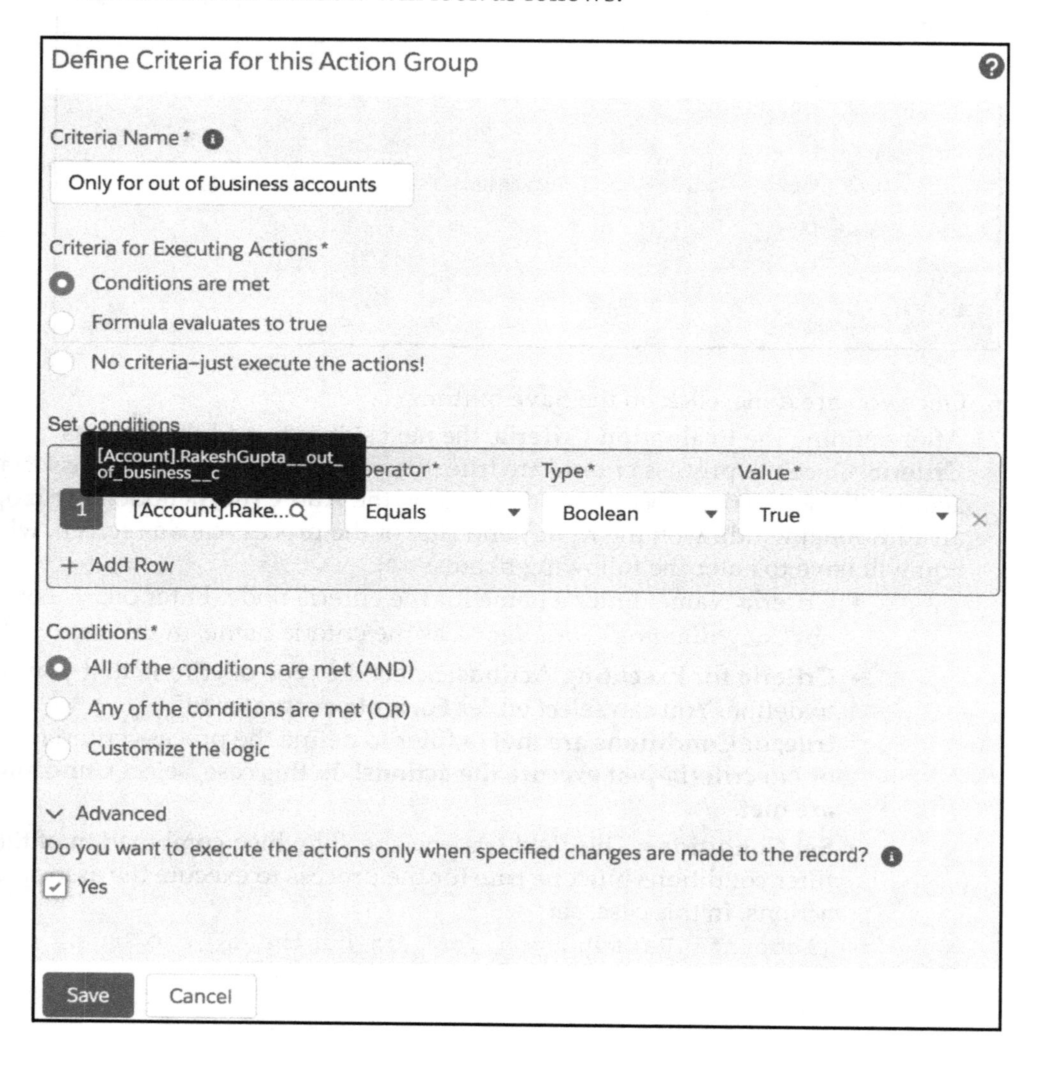

8. Once you are done, click on the **Save** button.
9. Once you are done with the process criteria node, the next step is to add an immediate action to delete open cases when the account is out of business. For this, we will use the **Apex** action, available in Process Builder. Click on **Add Action,** available under **IMMEDIATE ACTIONS**; it will open an additional window on the right-hand side of the process canvas screen, where you will have to enter the following details:
    - **Action Type**: Select the type of action; in this case, select **Apex**.
    - **Action Name**: Enter a name for this action - `Delete open cases`.
    - **Apex Class**: Select the Apex class that you want to execute; in this case, select `RakeshGupta__DeleteOpenCases`.
    - **Set Apex Variables**: Use this to set values for sObject variables and sObject list values. For the current use case, map the `AccountIds` field to `[Account].Id`.

The window will look as follows:

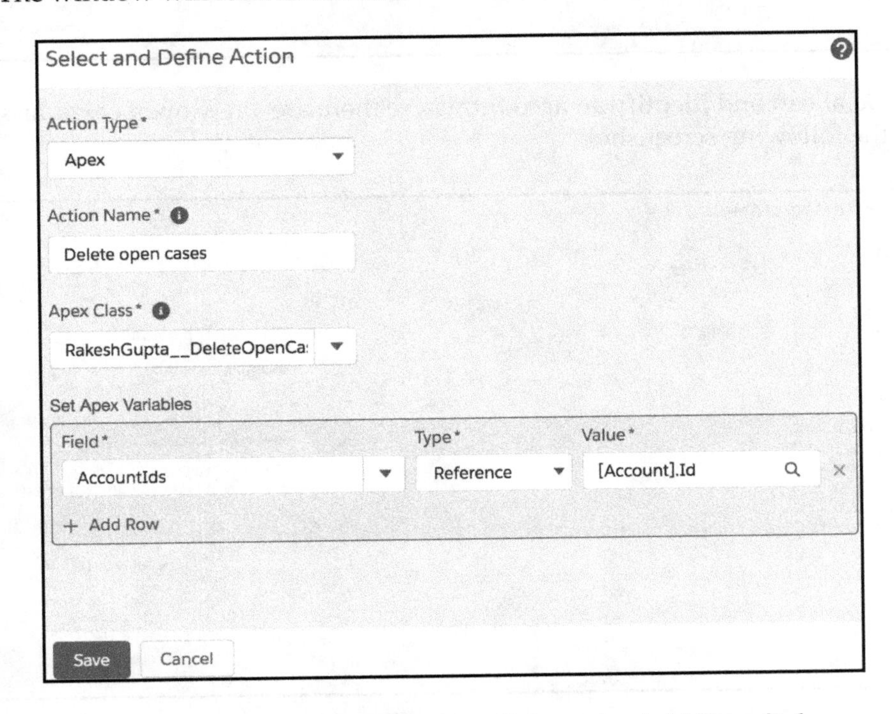

To assign values to multiple variables, click on the **Add Row** link.

10. Once you are done, click on the **Save** button.
11. The final step is to activate the process. Click on the **Activate** button on the button bar. Finally, the process will appear:

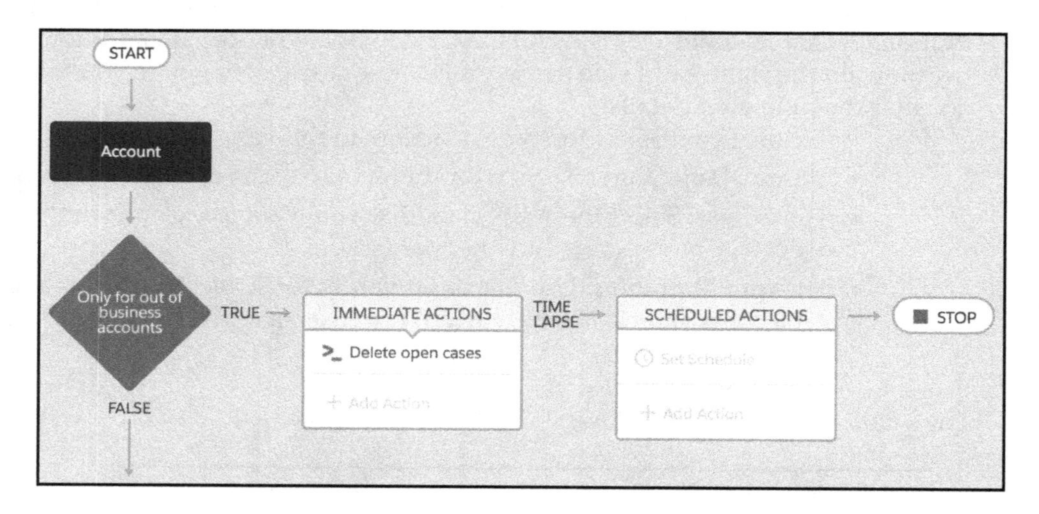

12. Go ahead and identify an account where there are a few open cases, as shown in the following screenshot:

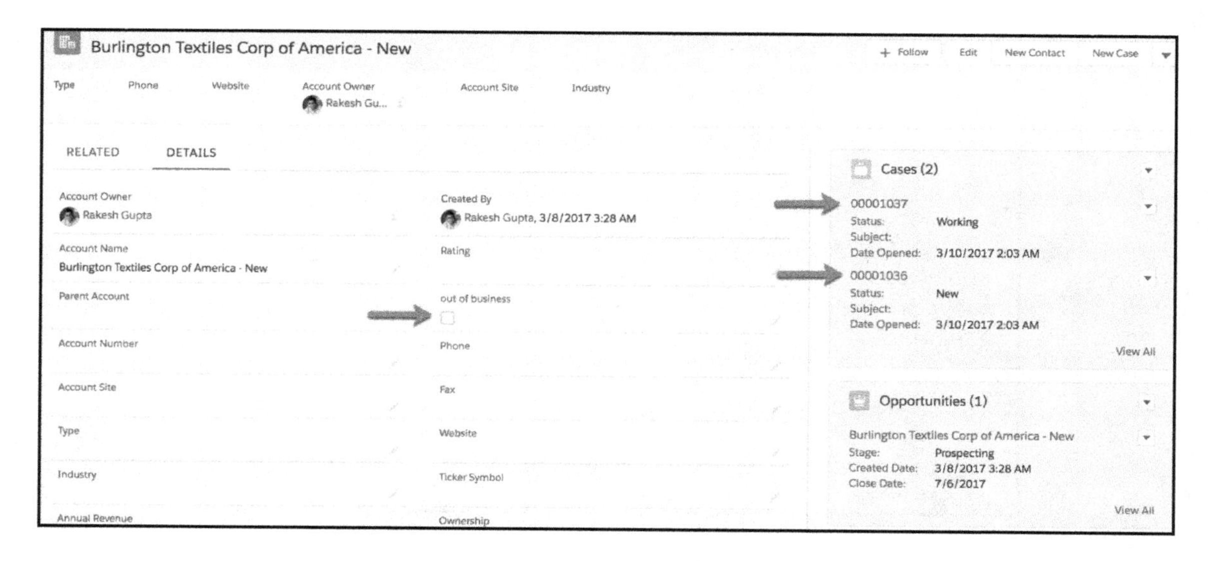

13. Now, update the **out of business** checkbox to **True,** and then reload the page. It will look like the following screenshot:

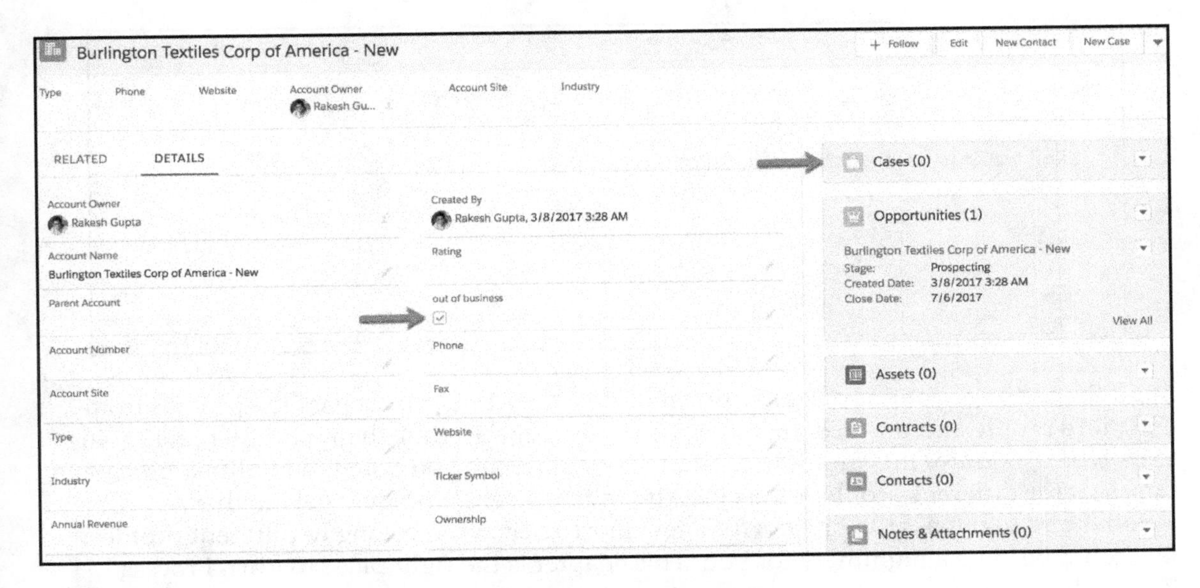

Before testing the process, make sure that you have activated it.

Hands on 5 – bypass processes using custom permissions

We have now created many processes using Process Builder. It is a quite easy and fun experience to create processes. These processes will execute as soon as they meet certain criteria. But there are some situations where a business may want to bypass these processes.

Let's look at an example. Suppose that Helina Jolly is working as a system administrator at Universal Containers. She developed the **Post Opportunity Information to Chatter Group** process in `Chapter 1`, *Getting Started with Lightning Process Builder*. She has now received a requirement to bypass this process for the system administrator.

It is quite easy to bypass the process for the system administrator, by adding an additional condition in your process, as shown in the following screenshot:

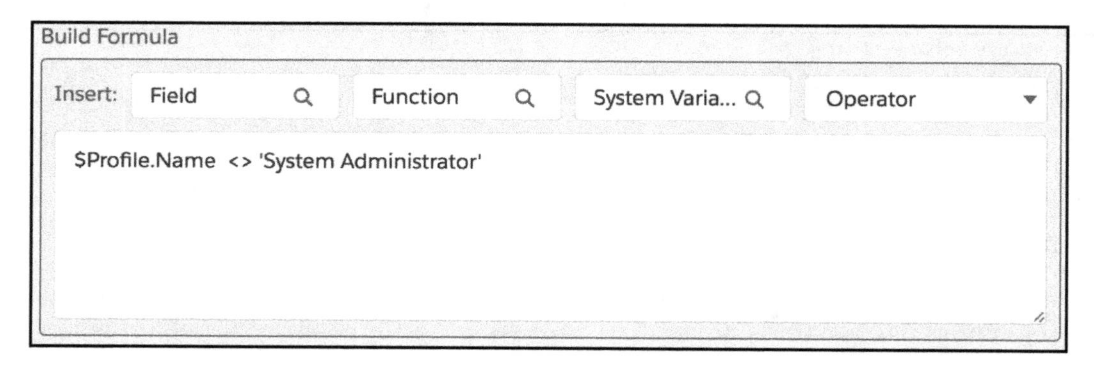

But if the business comes back after two months and asks her to bypass the process from one more profile and one user (belonging to a different profile), then the situation is going to be worse, because the question is: How many times are you going to modify a process to bypass it for different profiles or users? After a few weeks, you may get another requirement to bypass the process for a few more users, who belong to different profiles; this could become a nightmare for you. This chapter is the right place to learn Process Builder best practices.

Creating a custom permission

Let's look at a business scenario. Suppose that Helina Jolly is working as a system administrator at Universal Containers. She has received a requirement from the management to auto-update the related open opportunities status to **Closed Lost** if the **out of business** checkbox is checked on the account record. The management also wants to make sure that this process will not work for the system administrator and supply chain user profiles.

Using custom permissions, you can grant users access to custom apps. In Salesforce, you can use custom permissions to check which users can access a certain functionality. Custom permissions let you define access checks that can be assigned to users via permission sets or profiles, similar to how you assign user permissions and other access settings. We will now create a custom permission to bypass processes. Perform the following steps to create a custom permission:

1. Navigate to **Setup** (gear icon) | **Setup** | **PLATFORM TOOLS** | **Custom Code** | **Custom Permission** and click on the **New** button; it will redirect you to a new window, where you will have to enter the following details:
    - **Label**: Enter an easily identifiable term to recognize this custom permission. In this case, use **By Pass Process Builder**.
    - **Name**: This will be auto-populated, based on the **Label**.
    - **Description**: Write some meaningful text, so that other developers or administrators can easily understand why this custom permission was created.

    The **Custom Permission** page will look as follows:

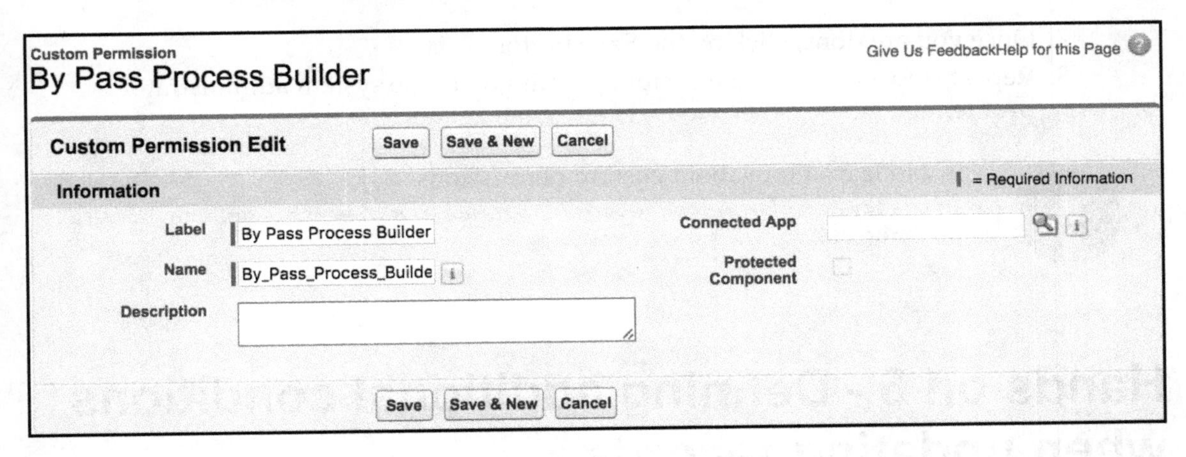

2. Once you are done, click on the **Save** button.
3. To assign a custom permission to a profile (you can also assign it to a permission set), navigate to **Setup** (gear icon) | **Setup** | **Users** | **Profiles,** and open the **Supply Chain User** profile (create a profile using the Salesforce license type, with the name **Supply Chain User,** if you haven't created it yet).
4. Navigate to **Apps** | **Custom Permission**.
5. Open the custom permission, and click on the **Edit** button.

6. Then, assign the **By Pass Process Builder** custom permission to the profile, as shown in the following screenshot:

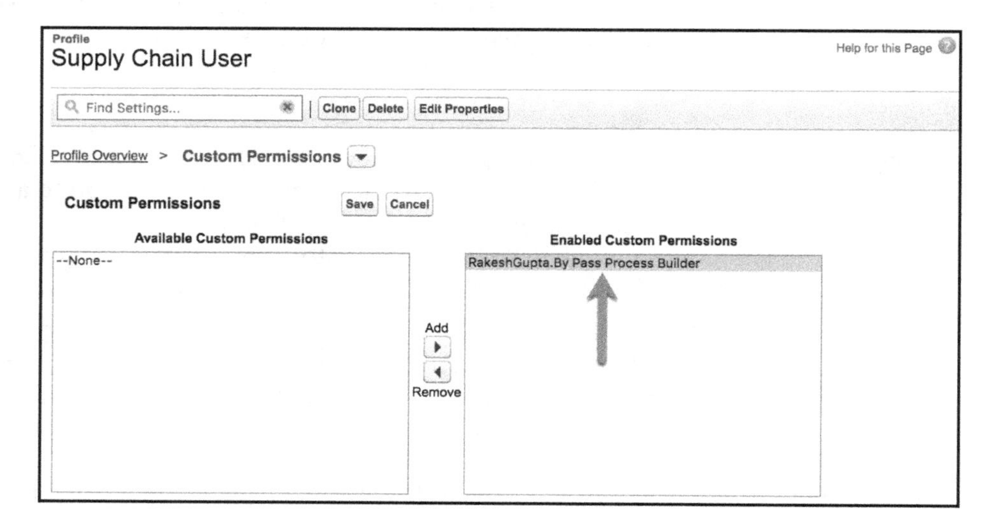

7. Once you are done, click on the **Save** button.
8. Repeat step 3 to assign the custom permission to the system administrator profile.

You can learn more about custom permissions at `https://help.salesforce.com/HTViewHelpDoc?id=changesets.htm&language=en_US`.

Hands on 6 - Defining additional conditions when updating records

When you update records using Process Builder, you can filter the records that you are updating using conditions. To add filter conditions to an update records action, there are two steps, as follows:

1. Select the **updated records meet all conditions** option
2. Set the conditions that you want to use to filter **updated records**

Now, we will create a process to fulfill the preceding business scenario and use custom permissions to bypass the process for the **System Administrator** and **Supply Chain User** profiles:

1. To create a process, navigate to **Setup** (gear icon) | **Setup** | **PLATFORM TOOLS** | **Process Automation** | **Process Builder**, click on the **New** button, and enter the following details:
    - **Name**: Enter the name of the process - **Update open opps - out of business accounts**.
    - **API Name**: This will be auto-populated, based on the name.
    - **Description**: Write some meaningful text, so that other developers or administrators can easily understand why this process was created.
    - **This process starts when**: Configure the process to start when a record is created or edited. In this case, select **A record changes**.
2. Once you are done, click on the **Save** button; it will redirect you to the process canvas, which allows you to create the process.
3. After **Define Process Properties**, the next task is to select the object upon which you want to create a process and define the **Evaluation Criteria**. For this, click on the **Add Object** node. It will open an additional window on the right-hand side of the process canvas screen, where you will have to enter the following details:
    - **Object**: Start typing, and then select the **Account** object.
    - **Start the process**: For **Start the process**, select **when a record is created or edited**. This means that the process will fire whenever a record gets created or edited.
    - **Recursion - Allow process to evaluate a record multiple times in a single transaction?**: Select this checkbox only when you want the process to evaluate the same record up to five times in a single transaction. In this case, leave this box unchecked.

The evaluation criteria page will look like the following screenshot:

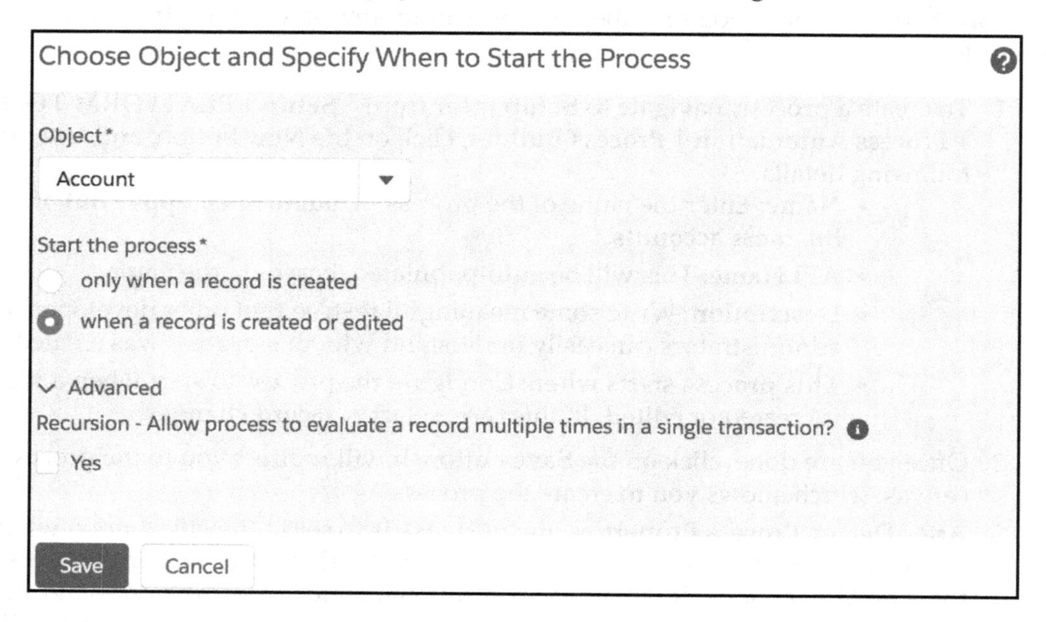

4. Once you are done, click on the **Save** button.
5. After defining the evaluation criteria, the next step is to add the **Process Criteria**. Once the process criteria are true, the process will execute the associated actions. To define the **Process Criteria**, click on the **Add Criteria** node. It will open an additional window on the right-hand side of the process canvas screen, where you will have to enter the following details:
    - **Criteria Name**: Enter a name for the criteria node - **Only for out of business accounts**.
    - **Criteria for Executing Actions**: Select the type of criteria that you want to define. You can select either **Formula evaluates to true,** or **Conditions are met** (a filter to define the process criteria), or **No criteria-just execute the actions!** In this case, select **Formula evaluates to true**.

- **Build Formula**: Use this to define the formula, using functions and fields. Select **System Variable**. It will open a popup, from which you can select the custom permission that you have created:

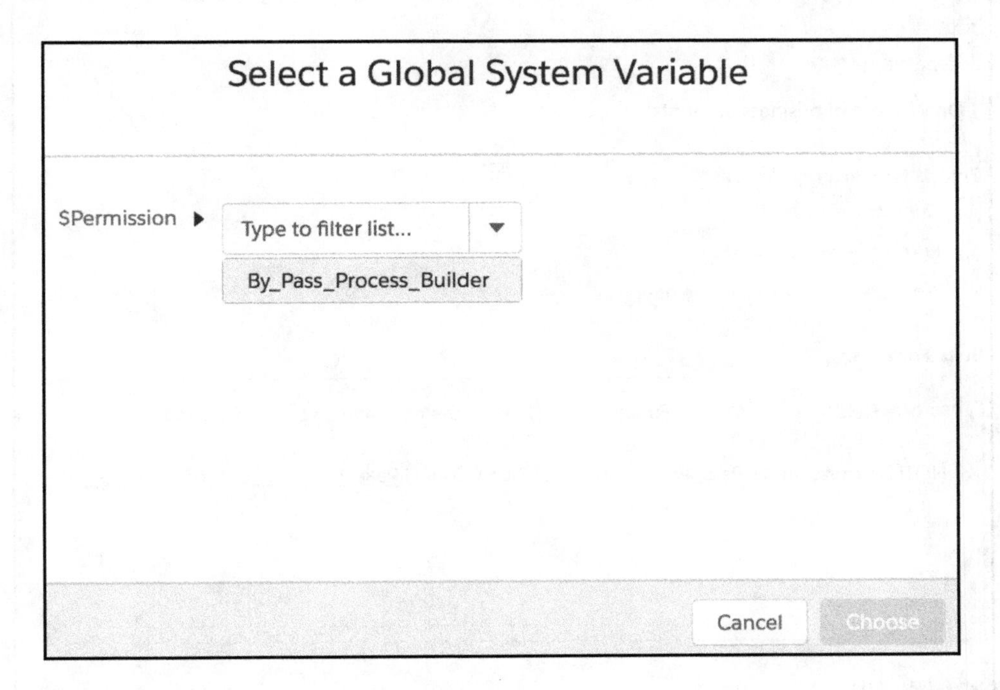

- Likewise, add an `[Account].RakeshGupta__out_of_business__c` field, as shown in the preceding screenshot.
- Under **Advanced**, select **Yes,** to execute the actions only when the specified changes are made.

The process criteria will look like the following screenshot:

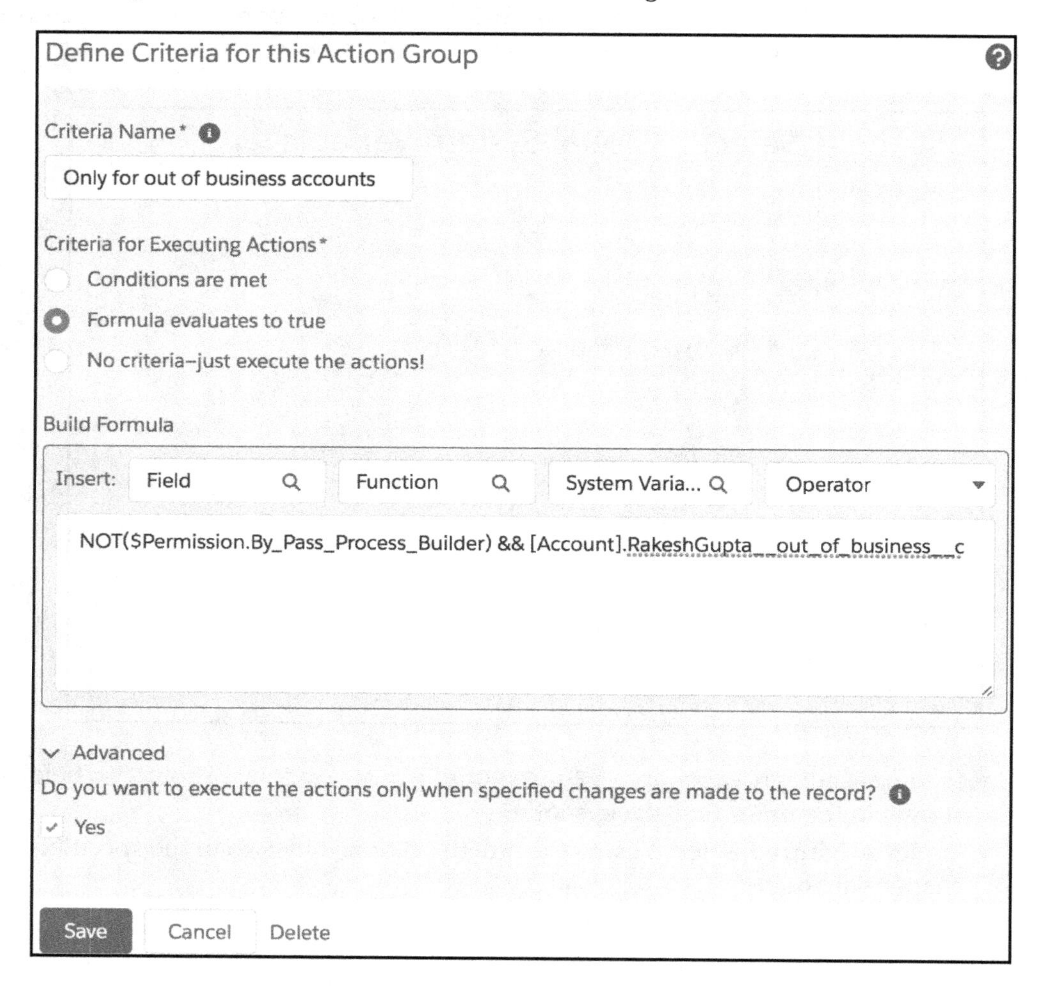

6. Once you are done, click on the **Save** button.
7. Once you are done with the **Process Criteria** node, the next step is to add an **Immediate Action,** to update open opportunities to Closed Lost. For this, we will use the **Update Records** action, available in Process Builder. Click on **Add Action,** under **IMMEDIATE ACTIONS**; it will open an additional window on the right-hand side of the process canvas screen, where you will have to enter the following details:
    - **Action Type**: Select the type of action; in this case, select **Update Records**.
    - **Action Name**: Enter a name for this action. Enter **Update open Opportunities** as the action name.
    - **Record Type**: Select the record (or records) that you need to update. Click on the **Record Type.** It will open a window, where you will see two options: **Select the Account record that started your process** and **Select a record related to the Account.** These are radio buttons, and only one can be selected to update the child record. In this case, choose to **Select a record related to the Account**. Then, in the **Type to filter list**, look for the **Opportunities** child object.
    - **Criteria for Updating Records**: Optionally, you can specify conditions to filter the records that you are updating. Select **Updated records meet all conditions** to filter out the related opportunities records. Then, set the **Stage** field value from **Does not equal** to **Closed Won**.
    - **Set new field values for the records you update**: Select the field whose value you want to set. In this case, set the **Stage** field to **Closed Lost**.

The preceding steps will look like the following screenshot:

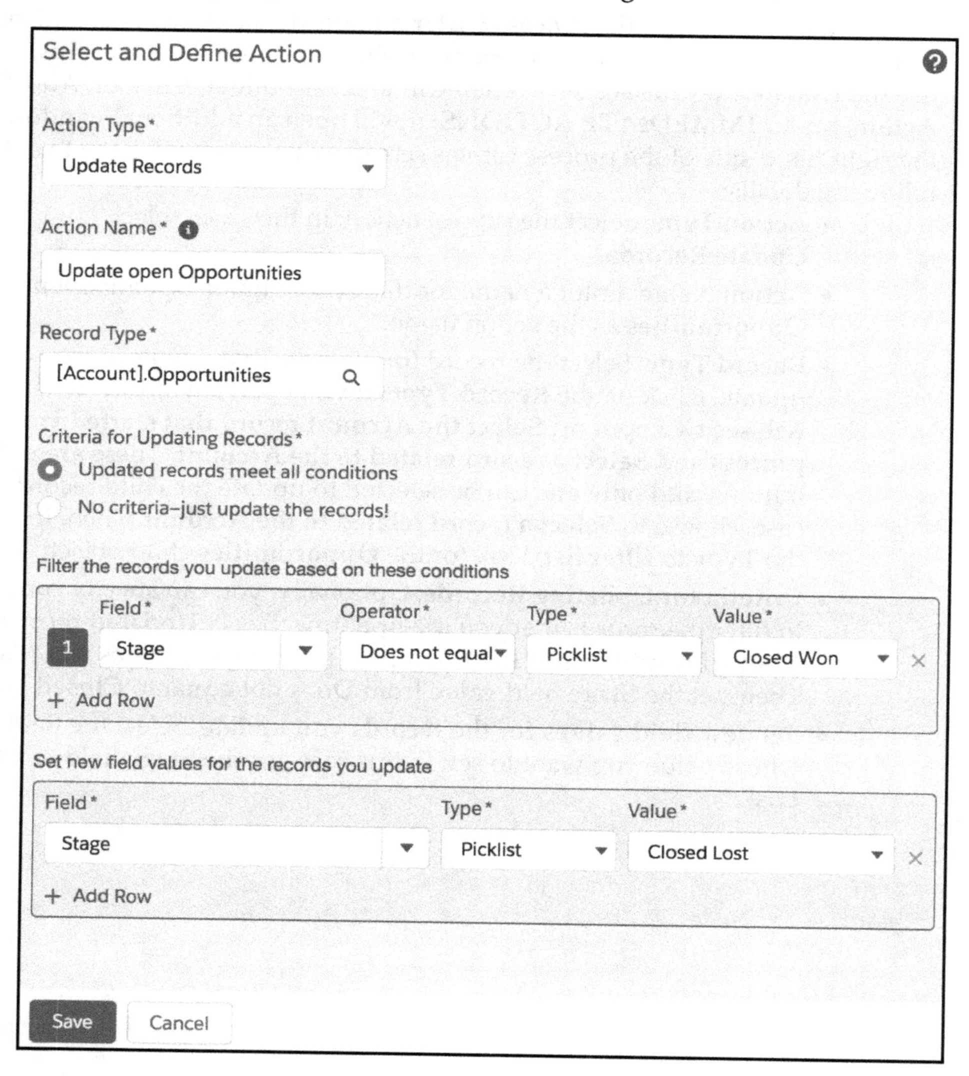

8. Once you are done, click on the **Save** button.
9. The final step is to activate the process. Click on the **Activate** button on the button bar. Finally, the process will appear, as shown in the following screenshot:

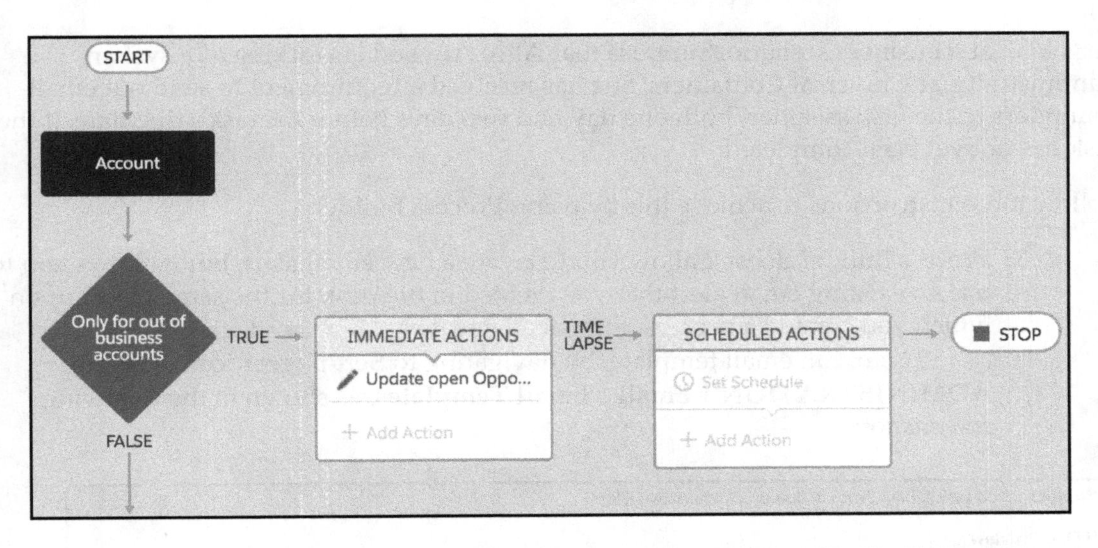

From now on, if an account becomes **out of business,** then the process will update the related open opportunities stage to Closed Lost. This process will only work when an account is updated by a user whose profile does not equal the system administrator or supply chain user (because we have assigned a custom permission to these profiles). If you want to bypass a process for a single user, consider using a permission set instead of a profile.

Hands on 7 – scheduling multiple groups of actions

With multiple schedules, you can easily optimize sales or support processes, automate follow-ups on outstanding cases and reminder notifications for a task, and incorporate all of your business requirements within a single process. For example, when the case **origin** is **Phone** and the **priority** is **high**, then you can execute multiple groups of scheduled actions, such as the following:

- Sending a reminder email to the case owner after one day, if a case is not closed.
- Sending a reminder email to the case owner and account owner after two days, if a case is still open.
- Sending a satisfaction survey email to case contacts, two days after the case closure.

Let's look at a business scenario. Suppose that Alice Atwood is working as a system administrator at Universal Containers. She has received a requirement to send out email reminders to the task assignee, both one day and two days before the task's due date, if the task has not yet been completed.

Follow these instructions to achieve this by using Process Builder:

1. Process Builder doesn't allow you to create a new email alert, but it allows you to use an existing email alert that you created in the past, for the same object upon which you created a process. First of all, create a `Task reminder notification` email template, by navigating to **Setup** (gear icon) | **Setup** | **ADMINISTRATION** | **Email** | **Email Templates**, as shown in the following screenshot:

Subject You task {!Task.Who} is due on {!Task.ActivityDate}

HTML Preview

Hi There,

You task {!Task.Who} is due on {!Task.ActivityDate}. Below is some key information for you

Assigned By :- {!Task.CreatedBy}
Related To :- {!Task.What}
Due Date :- {!Task.ActivityDate}

Best Regards,
Universal Container Sales Team

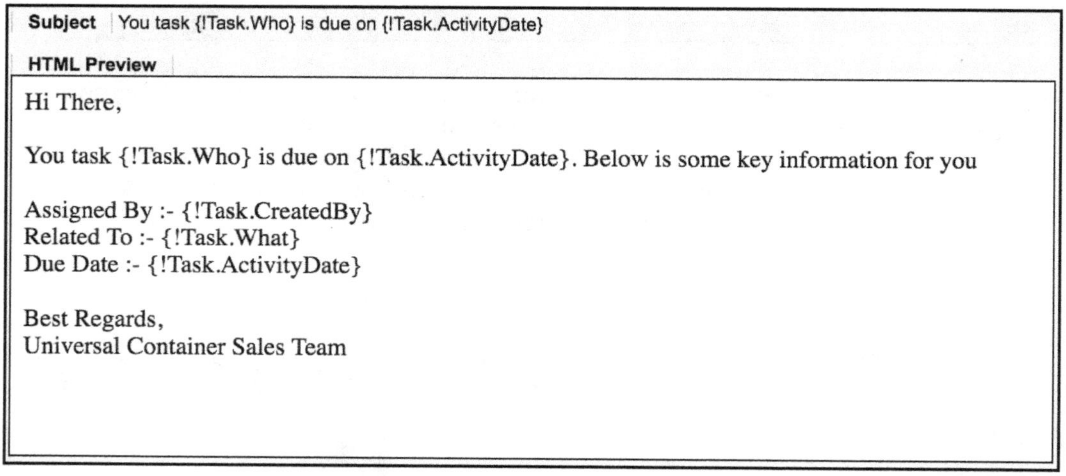

3. The second step is to create an email alert on the **Opportunity** object, by navigating to **Setup** (gear icon) | **Setup** | **PLATFORM TOOLS** | **Process Automation** | **Process Builder** | **Workflow Actions** | **Email Alerts**. Click on the **New Email Alert** button, and save it with the name **Email to Task assignee**. It should look like the following screenshot:

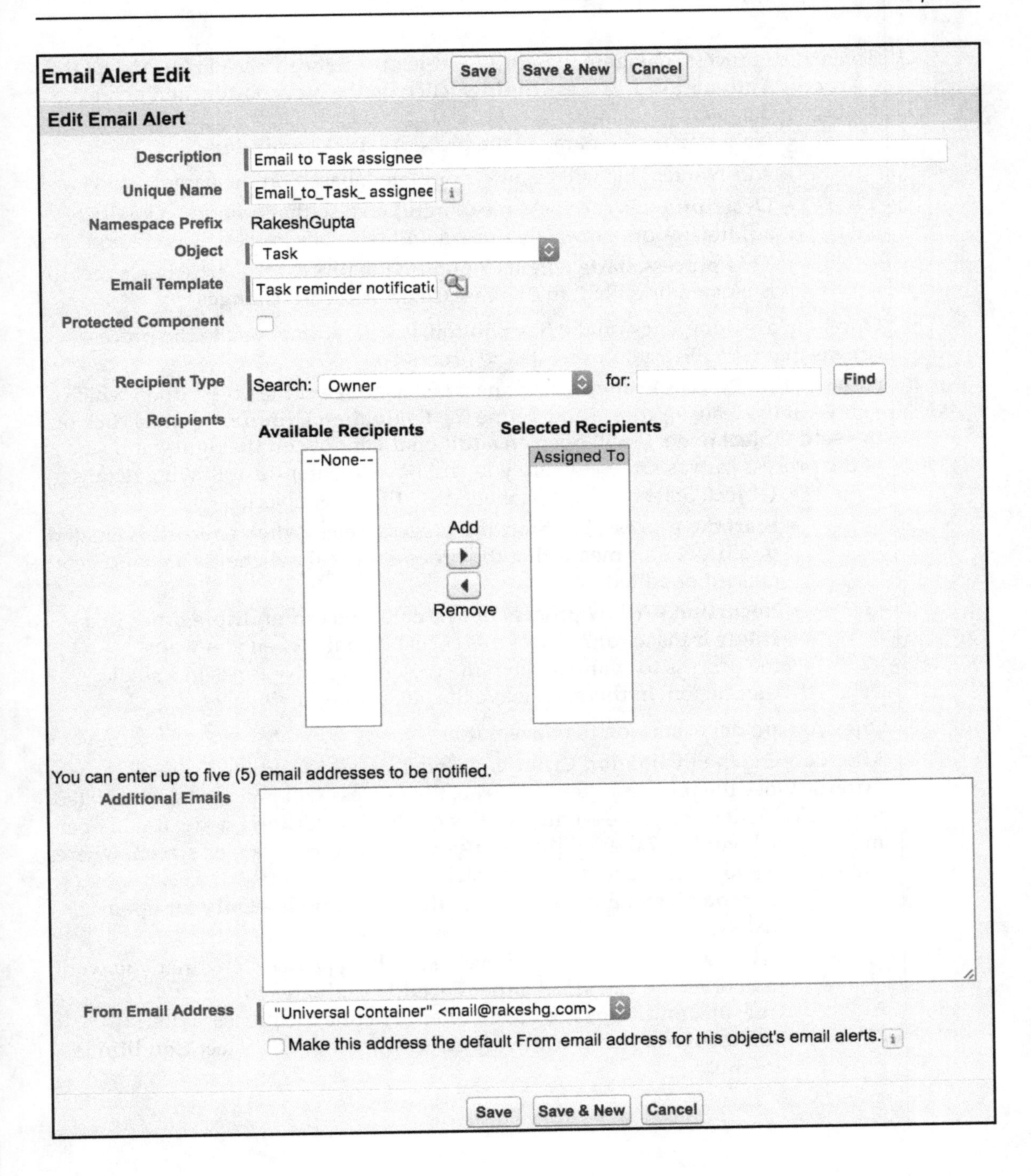
Email Alert Edit
Save
Save & New
Cancel
Edit Email Alert
Description
Email to Task assignee
Unique Name
Email_to_Task_ assignee
Namespace Prefix
RakeshGupta
Object
Task
Email Template
Task reminder notificati
Protected Component
Recipient Type
Search:
Owner
for:
Find
Recipients
Available Recipients
--None--
Selected Recipients
Assigned To
Add
Remove
You can enter up to five (5) email addresses to be notified.
Additional Emails
From Email Address
"Universal Container" <mail@rakeshg.com>
Make this address the default From email address for this object's email alerts.
Save
Save & New
Cancel

4. To create a process, navigate to **Setup** (gear icon) | **Setup** | **PLATFORM TOOLS** | **Process Automation** | **Process Builder**, click on the **New** button, and enter the following details:
    - **Name**: Enter the name of the process – **Task reminders**.
    - **API Name**: This will be auto-populated, based on the name.
    - **Description**: Write some meaningful text, so that other developers or administrators can easily understand why this process was created.
    - **This process starts when**: Configure the process to start when a record is created or edited. In this case, select **A record changes**.
5. Once you are done, click on the **Save** button; it will redirect you to the process canvas that will allow you to create the process.
6. After **Define Process Properties**, the next task is to select the object upon which you want to create a process and define the **Evaluation Criteria**. For this, click on the **Add Object** node. It will open an additional window on the right-hand side of the process canvas screen, where you will have to enter the following details:
    - **Object**: Start typing, and then select the **Task** object.
    - **Start the process**: For **Start the process**, select **when a record is created or edited**. This means that the process will fire whenever a record gets created or edited.
    - **Recursion - Allow process to evaluate a record multiple times in a single transaction?**: Select this checkbox only when you want the process to evaluate the same record up to five times in a single transaction. In this case, leave the box unchecked.
7. Once you are done, click on the **Save** button.
8. After defining the **Evaluation Criteria**, the next step is to add the **Process Criteria**. Once the process criteria are true, the process will execute the associated actions. To define the process criteria, click on the **Add Criteria** node. It will open an additional window on the right-hand side of the process canvas screen, where you will have to enter the following details:
    - **Criteria Name**: Enter a name for the criteria node - **Only for open tasks**.
    - **Criteria for Executing Actions**: Select the type of criteria that you want to define. You can select either **Formula evaluates to true**, or **Conditions are met** (a filter to define the process criteria), or **No criteria-just execute the actions!** In this case, select **Conditions are met**.

- **Set Conditions**: This field lets you specify which combination of the filter conditions must be true for the process to execute the associated actions. In this case, select **[Task].Status Does not equal Completed.**
- **Conditions**: In the **Conditions** section, select **All of the conditions are met (AND)**. This field lets you specify which combination of the filter conditions must be true for the process to execute the associated actions.
- Under **Advanced**, select **Yes,** to execute the actions only when the specified changes are made.

The preceding steps will look as follows:

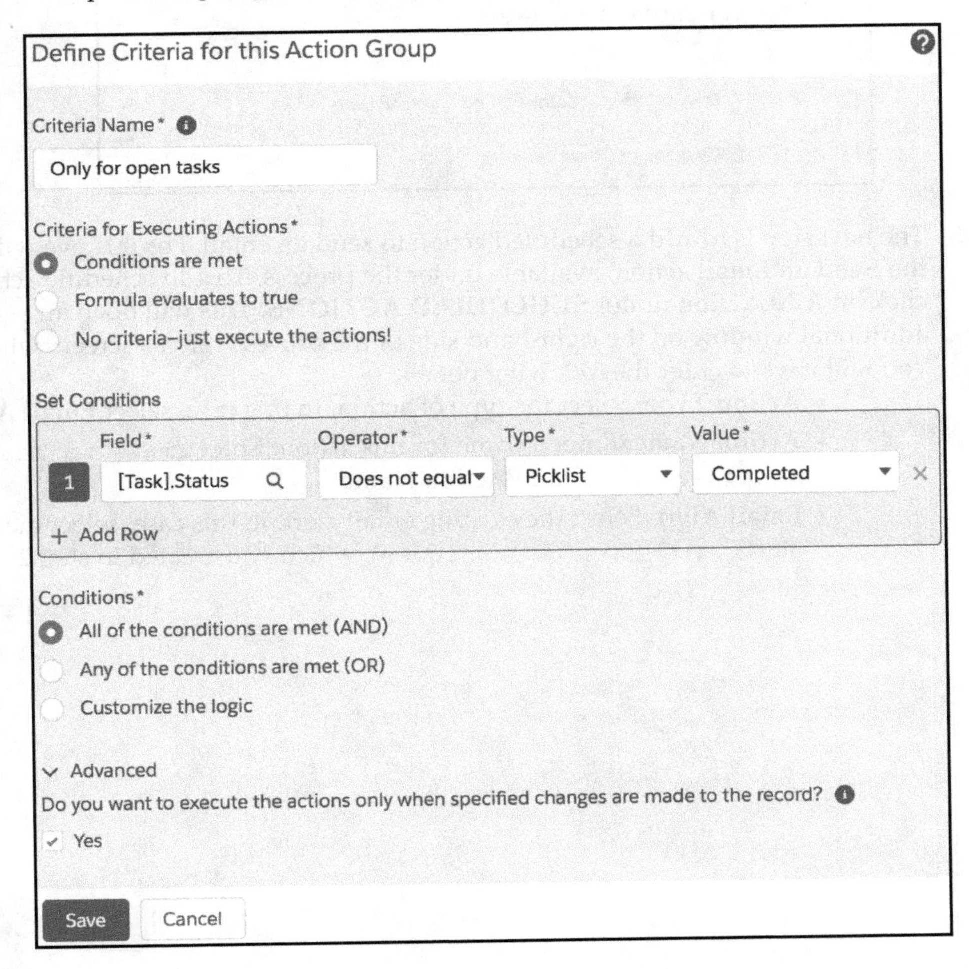

In the very first step, we have selected **when a record is created or edited**; if you want to create scheduled actions, make sure to select the **Do you want to execute the actions only when specified changes are made to the record?** checkbox. This is similar to **Evaluate the rule when a record is created, and any time it's edited to subsequently meet criteria** in the **Workflow Rule**.

8. Once you are done, click on the **Save** button.
9. To add the scheduled time, click on **Set Schedule,** available under **SCHEDULED ACTIONS**, and set **Set Time for Action to Execute** to **2 Days Before ActivityDate** (the task due date), as shown in the following screenshot:

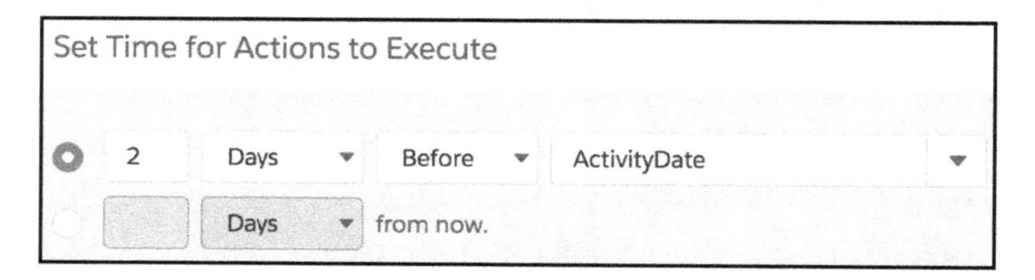

10. The next step is to add a scheduled action to send an email. For this, we will use the **Send an Email** action, available under the process. To add schedule actions, click on **Add Action** under **SCHEDULED ACTIONS**. This will open an additional window on the right-hand side of the process canvas screen, where you will have to enter the following details:
    - **Action Type**: Select the type of action. In this case, select **Email Alerts**.
    - **Action Name**: Enter a name for this action. Enter `Email to Task Assignee- 2 Days` in **Action Name**.
    - **Email Alert**: Select the existing email alert. In this case, select the email alert (`Email_to_Task_assignee`), which you created in step 2.

The preceding steps will appear as follows:

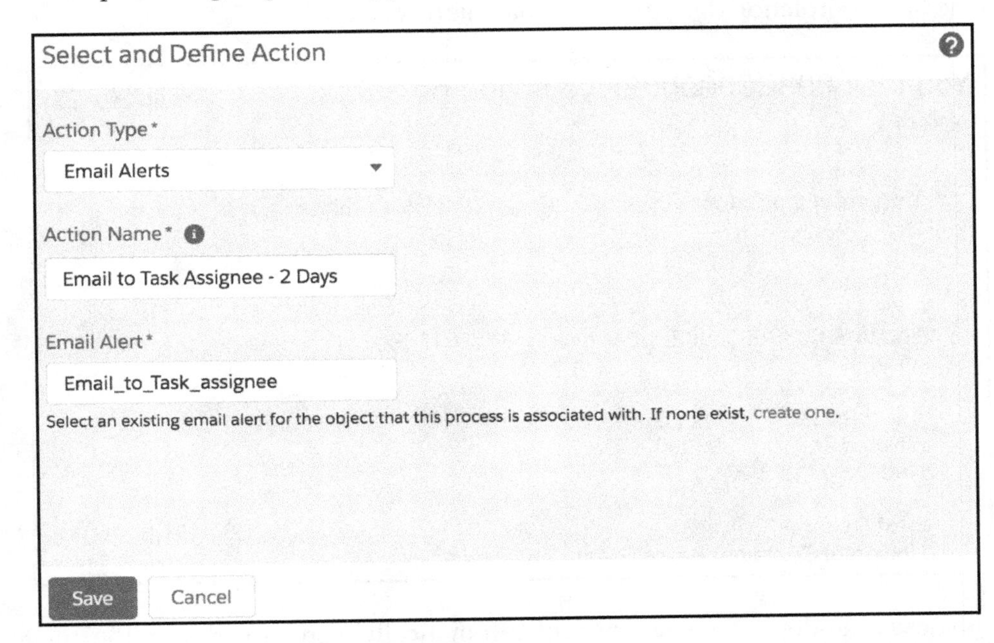

11. Once you are done, click on the **Save** button.
12. Repeat steps 9, 10, and 11, to add one more scheduled action for one day before.
13. Once you are done with the process creation, the final step is to activate it. To activate the process, click on the **Activate** button on the button bar. Finally, the process will appear, as shown in the following screenshot:

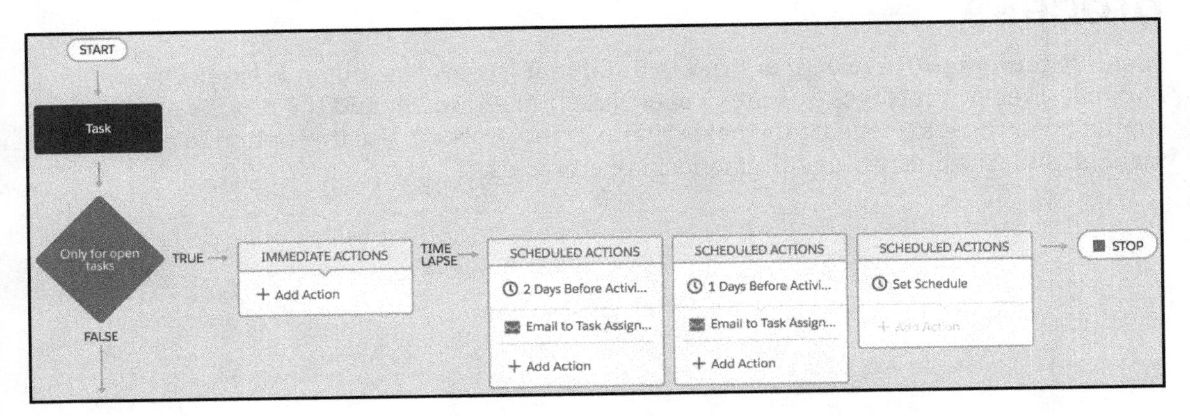

From now on, this process will send reminder emails to the task assignee one and two days prior to the task completion date, and the email alert will look as follows:

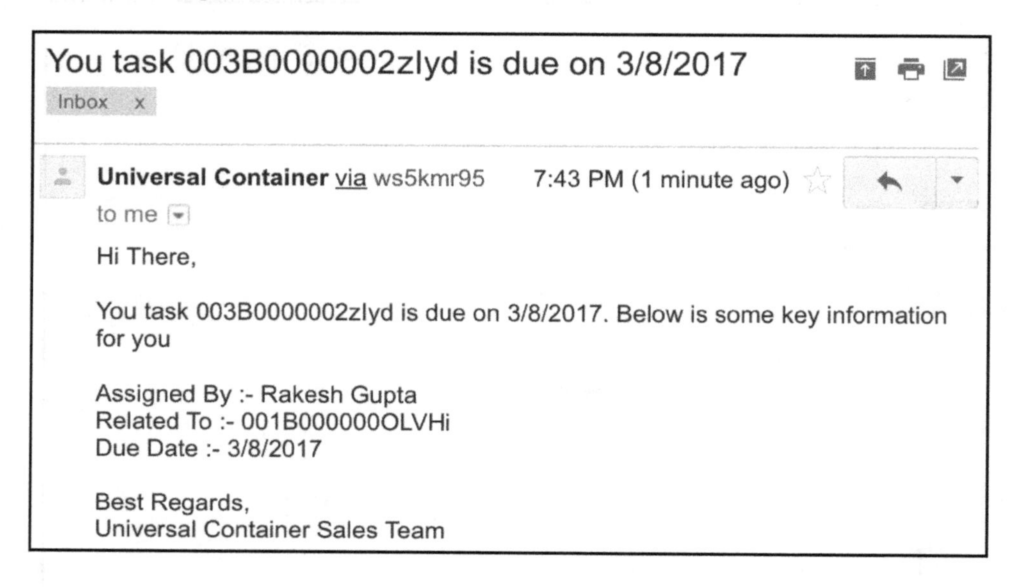

These time-dependent actions will automatically be removed from the queue if the assignee closes a task before the due date.

Hands on 8 – executing multiple criteria of a process

Since the Summer 2016 release of Process Builder, it has become possible to choose what happens after your process executes a specific action group. Should the process stop, or should it continue to evaluate the next criteria in the process? Use this option to efficiently manage multiple business requirements in one process.

It is easy to manage all of your processes for a given object, like a lead, in one place:

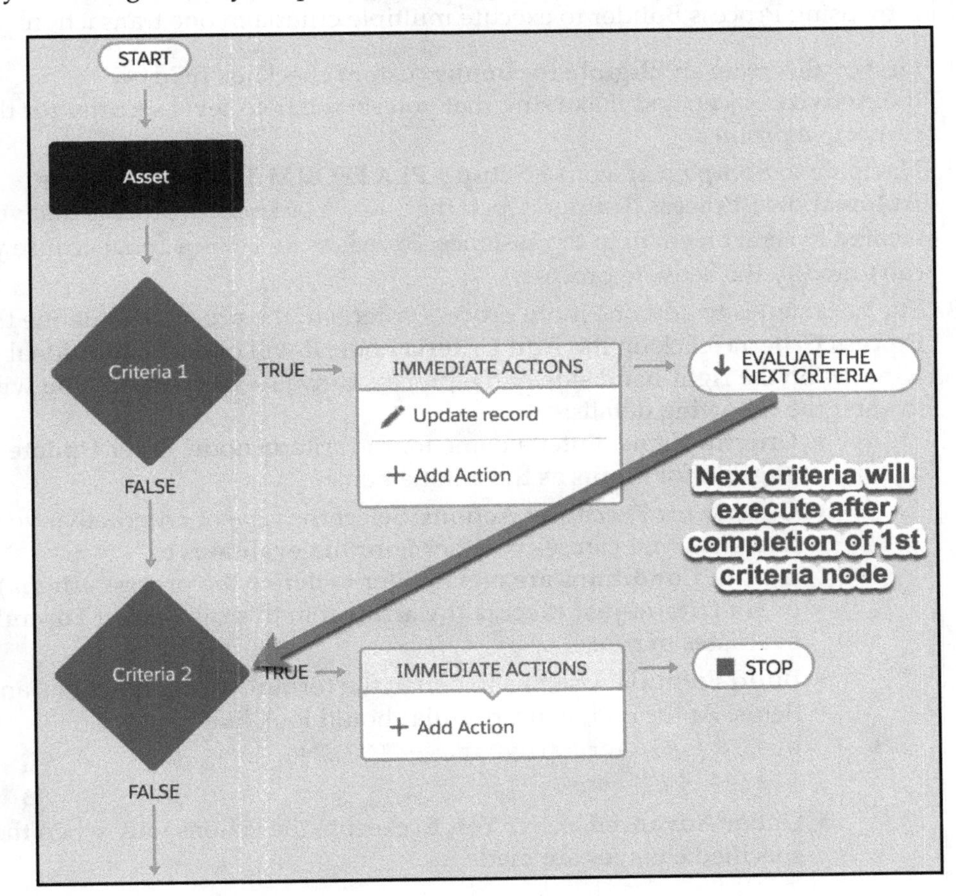

Let's look at a business scenario. Suppose that Alice Atwood is working as a system administrator at Universal Containers. She has received a requirement from the management; once a task has successfully closed five days before its due date, auto-update the **Eligible for bonus** checkbox to true.

There are two possible solutions for the preceding business requirement:

- Create a new process to solve the business requirement
- Use an existing process, and add one more criteria to it

We will use the second approach to solve the preceding business requirement. Follow these instructions by using Process Builder to execute multiple criteria in one transaction:

1. First of all, create an **Eligible for bonus** custom checkbox field on the **Activity** object, and make sure that you set the field-level security for the respective profiles.
2. Navigate to **Setup** (gear icon) | **Setup** | **PLATFORM TOOLS** | **Process Automation** | **Process Builder**. Open the `Task reminders` process that you created to send an email to the assignee. Save it as `New Version`, because you can't modify the activate process.
3. The next step is to add one more process criteria to the process. To define the **Process Criteria**, click on the **Add Criteria** node; it will open an additional window on the right-hand side of the process canvas screen, where you will have to enter the following details:
    - **Criteria Name**: Enter a name for the criteria node. Enter **Update eligible for bonus** as the criteria name.
    - **Criteria for Executing Actions**: Select the type of criteria that you want to define. You can select either **Formula evaluates to true,** or **Conditions are met** (a filter to define the process criteria), or **No criteria-just execute the actions!** In this case, select **Formula evaluates to true.**
    - **Build Formula**: Use this to define the formula using functions and fields. At the end, your formula should look like this: `[Task].ActivityDate - TODAY() >=5 && [Task].IsClosed`.
    - Under **Advanced**, select **Yes,** to execute the actions only when the specified changes are made.

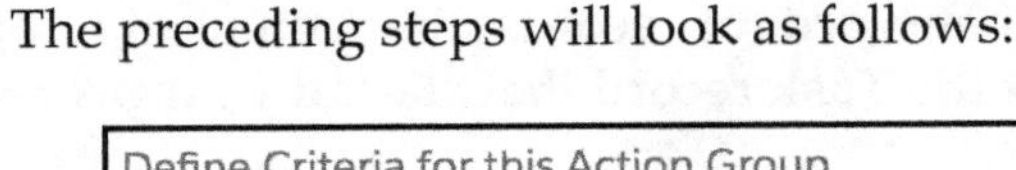

The preceding steps will look as follows:

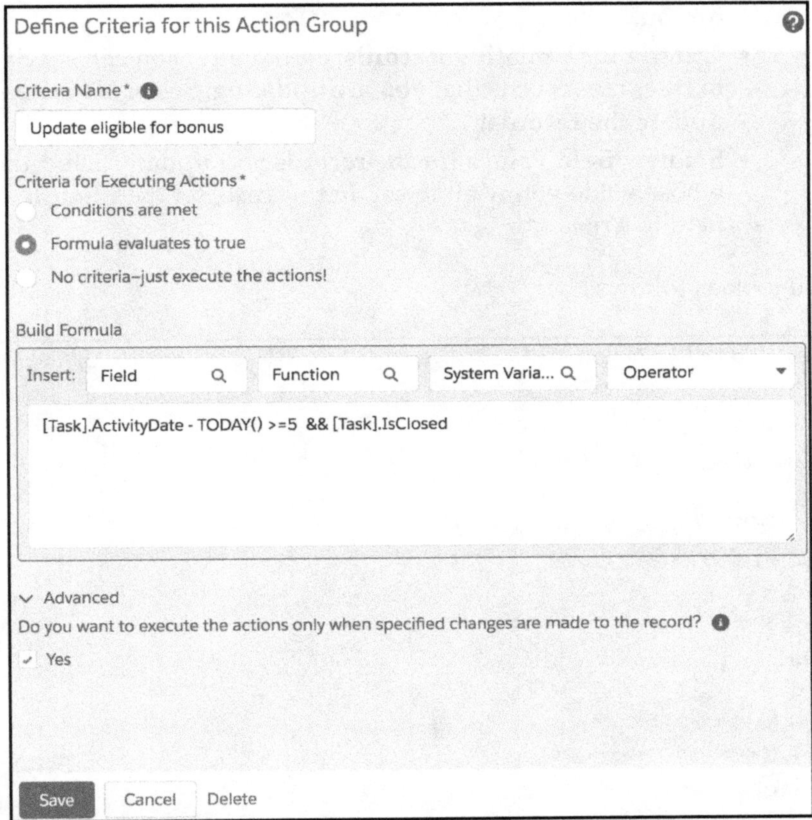

4. Once you are done, click on the **Save** button.
5. Once you are done with the process criteria node, the next step is to add an immediate action to update the task field **eligible for bonus** to true, if the task is closed five days before the due date. For this, we will use the **Update Records** action, available in Process Builder. Click on **Add Action,** available under **IMMEDIATE ACTIONS**; it will open an additional window on the right-hand side of the process canvas screen, where you will have to enter the following details:
    - **Action Type**: Select the type of action; in this case, select **Update Records**.
    - **Action Name**: Enter a name for this action. Enter **Update Eligible for bonus to true** as the action name.

- **Record Type**: Select the record (or records) that you need to update. In this case, select **Select the Task record that started your process option.**
- **Criteria for Updating Records**: Optionally, you can specify conditions to filter the records that you are updating. Select **No criteria—just update the records!**
- **Set new field values for the records you update**: Select the field whose value you want to set. In this case, set the **Eligible for bonus** field to **True.**

It will look as follows:

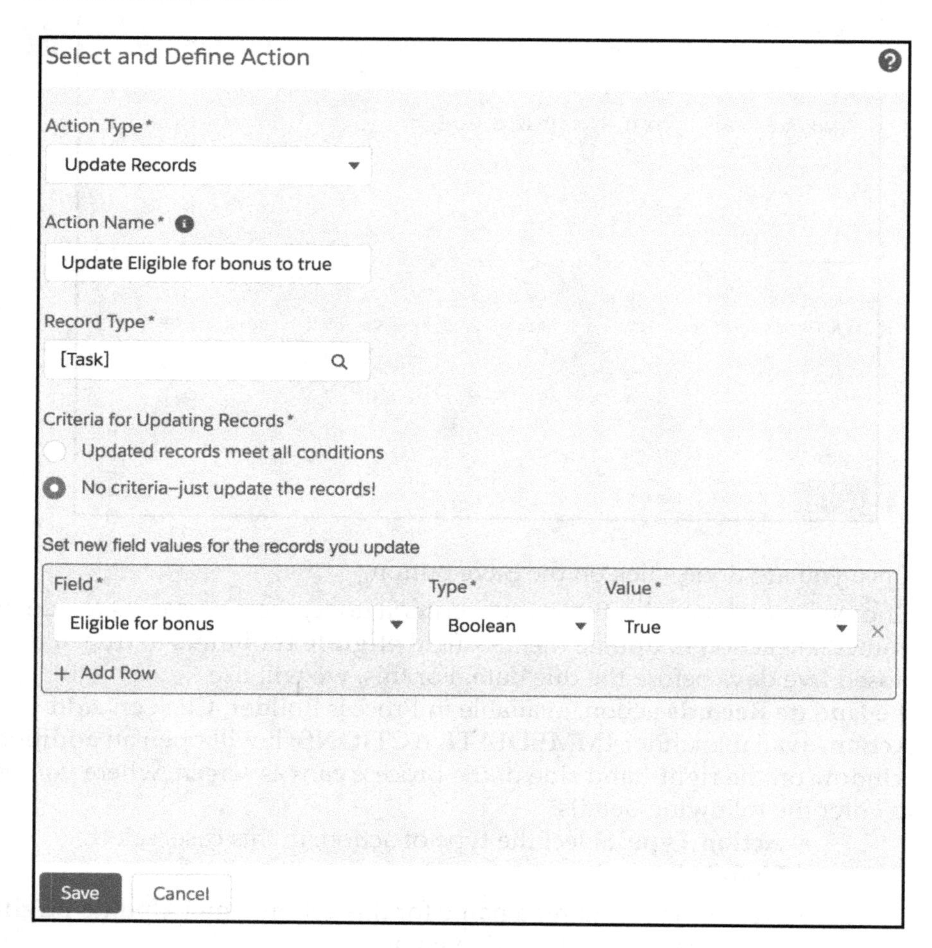

6. Once you are done, click on the **Save** button.
7. The final step is to activate the process. Click on the **Activate** button, available on the button bar. Finally, the process will appear, as shown in the following screenshot:

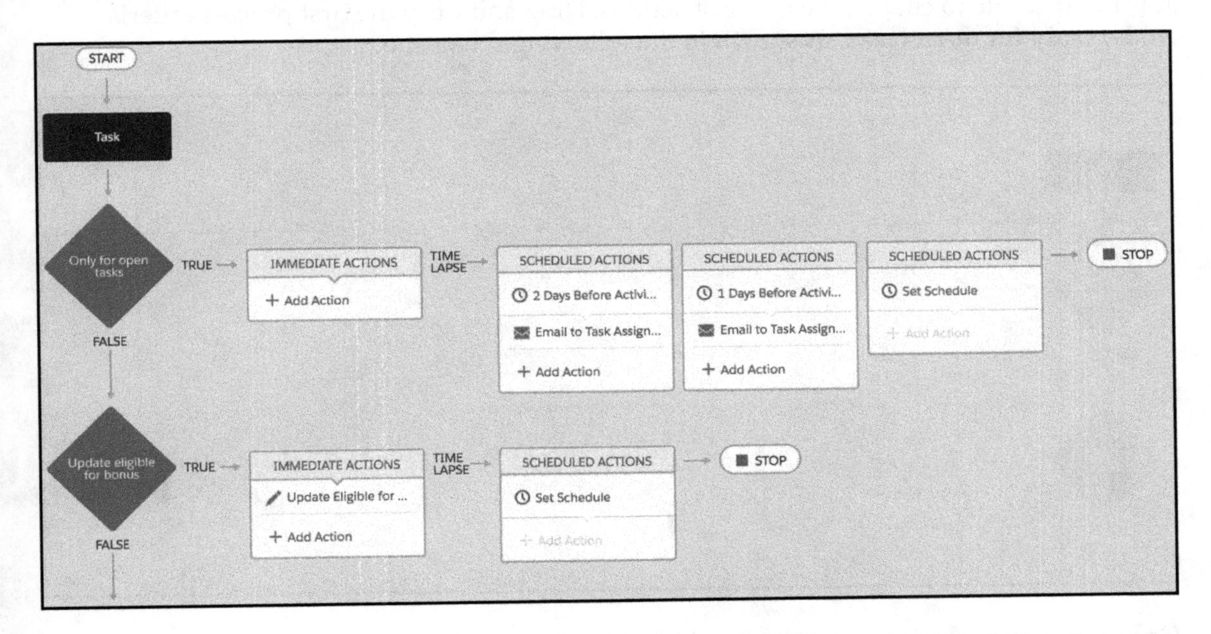

Still, only one criteria node will execute in one transaction, because the process works on the if...else statement. This means that only one criteria node can be true in a transaction.

Reordering the criteria node in Process Builder

Before executing both of the criteria in a transaction, first, let's look at how to reorder the criteria nodes. You can reorder the criteria nodes by just dragging and dropping them, but it is not possible to change the order of actions. Drag and drop the first process criteria node, **Only for open tasks**, as shown in the following screenshot:

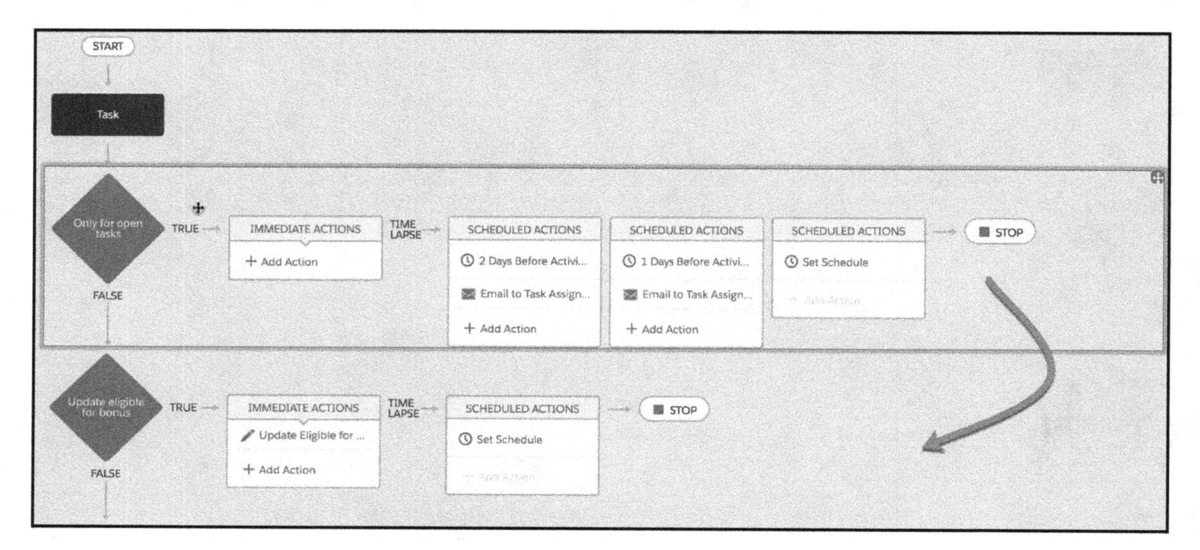

Criteria are evaluated in the order in which they shown on the process canvas. When the criteria are true, the process executes the associated action group and stops evaluating additional criteria. After reordering your process, it should look as follows:

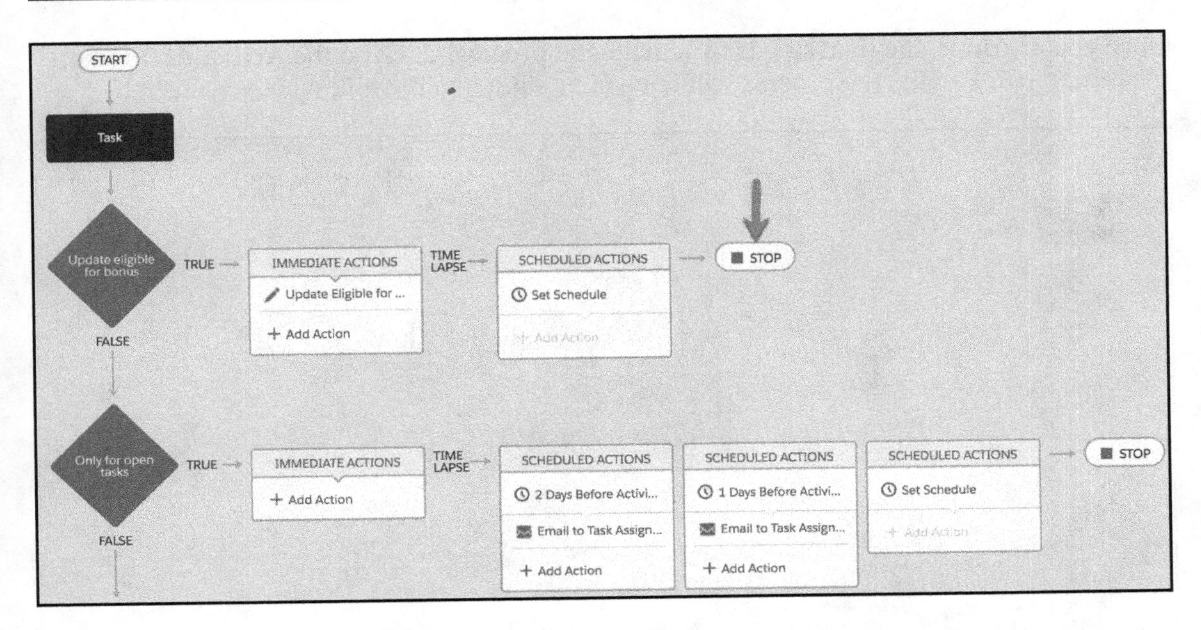

If you want to execute multiple criteria nodes in one transaction, you have to connect both of the criteria. To do that, click on **STOP**, as shown in the preceding screenshot. It will open an additional window on the right-hand side of the process canvas screen, where you will find following options:

- **Stop the process**: You will be able to stop the process after executing the actions. By default, each action group is set to stop after executing actions.
- **Evaluate the next criteria**: Select this if you want to continue evaluating the next defined criteria in the same transaction.

In this case, select **Evaluate the next criteria**, as shown in the following screenshot:

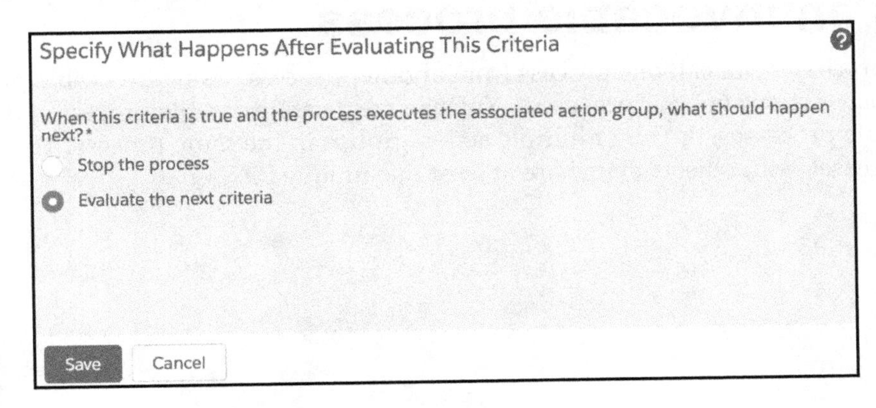

Once you are done, the final step is to activate the process. Click on the **Activate** button on the button bar. Finally, the process will appear, as shown in the following screenshot:

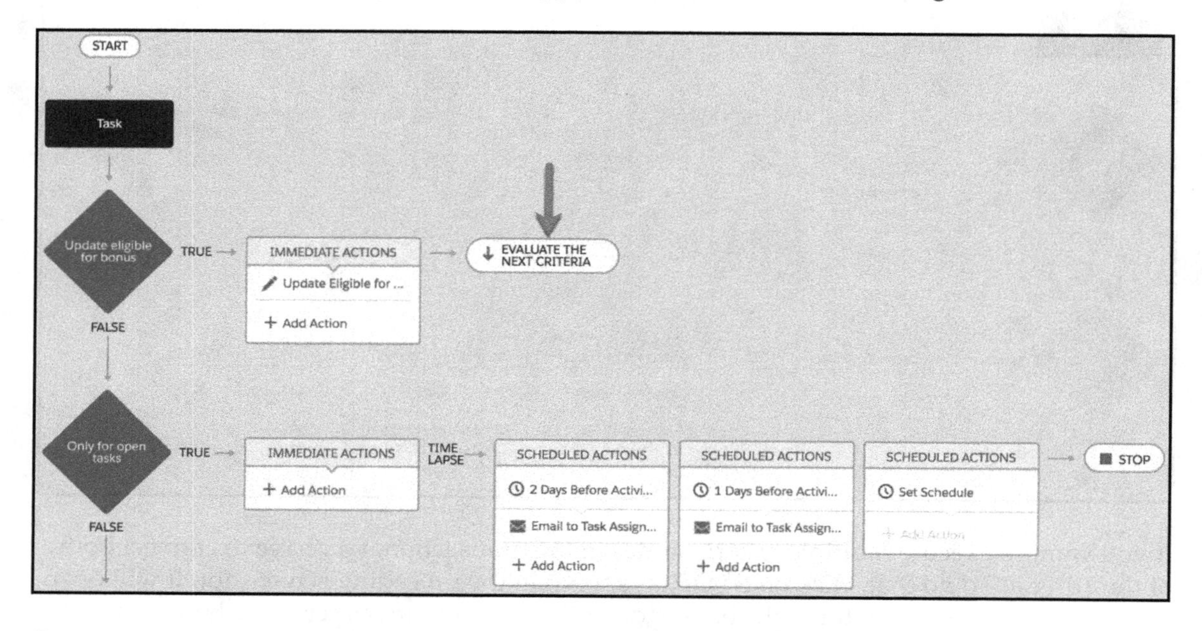

From now on, the process will send reminder emails to the assignee, one and two days prior to the task's due date; it will update **Eligible for bonus** to **True** if a task has successfully closed five days before the due date.

Hands on 9 – creating reusable processes using an invocable process

Invocable processes can call one process from another process. Using invocable processes, you can reuse sections of your processes. You can create an invocable process and call it from multiple processes, or from multiple action groups in the same process. You can invoke processes with objects that share at least one unique ID.

For example, in the Opportunity and Quote objects, the `OpportunityID` field is unique to Opportunity, and is also used by Quote. You can create an invocable process that updates an Opportunity record. Then, you can invoke it from the following:

- A process that updates an Opportunity record
- A process that updates a Quote record

When you create a process, make sure that you start it when another process invokes it, by selecting **It's invoked by another process.**

Let's look at a business scenario. Suppose that Alice Atwood is working as a system administrator at Universal Containers. She has received a requirement from the management to automate the sales process a bit; that is, if a quote is denied by a customer, then a process should auto-update the opportunity to Closed Lost.

There are numerous methods to solve the preceding business requirement, including the following:

- Process Builder
- Visual Workflow and Process Builder
- Visual Workflow and inline Visualforce page
- Apex trigger

We will use an invocable process to solve this business requirement. We will have to create two processes for this, as follows:

- The first process (**It's invoked by another process**) is placed on the **Opportunity** object, to update the **Stage** to **Closed Lost**.
- Another process (**A record changes**) is placed on the **Quote** object, and it will fire only when the quote **status** is updated to **Denied**.

The benefit of using this approach is that if you get a future requirement to update the opportunities stage to Closed Lost when the **account Active** field is updated to `false`, then you can call this process from your account process without adding another record update.

Perform the following steps to solve the preceding business requirement:

1. Navigate to **Setup** (gear icon) | **Setup** | **PLATFORM TOOLS** | **Process Automation** | **Process Builder**, click on the **New** button, and enter the following details:
    - **Name**: Enter the name of the process -**Update Opportunity Stage.**
    - **API Name**: This will be auto-populated, based on the name.

- **Description**: Write some meaningful text, so that other developers or administrators can easily understand why this process was created.
- **This process starts when**: Configure the process to start when a record is created or edited. In this case, select **It's invoked by another process**.

2. Once you are done, click on the **Save** button. It will redirect you to the process canvas, which will allow you to create or modify the process.
3. After **Define Process Properties**, the next task is to select the object upon which you want to create a process and define **Evaluation Criteria**. For this, click on the **Add Object** node. It will open an additional window on the right-hand side of the process canvas screen, where you will have to enter the **Object**. Start typing, and then select the **Opportunity** object. A window will appear, as shown in the following screenshot:

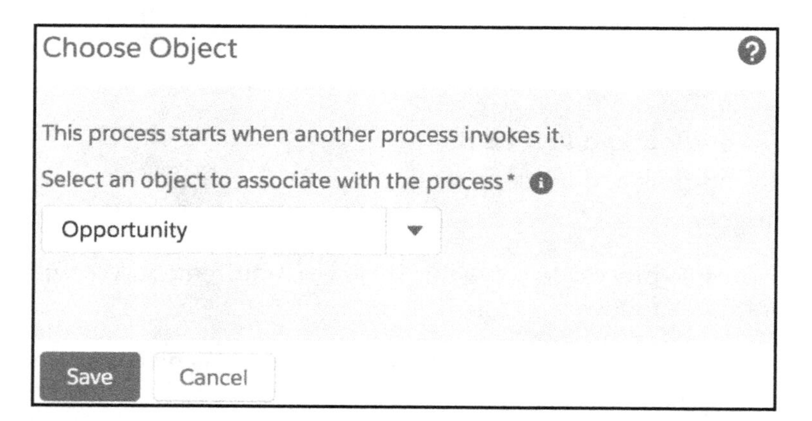

4. Once you are done, click on the **Save** button.
5. After defining the **Evaluation Criteria**, the next step is to add the **Process Criteria**. Once the process criteria are true, the process will execute the associated actions. To define the process criteria, click on the **Add Criteria** node. It will open an additional window on the right-hand side of the process canvas screen, where you will have to enter the following details:
    - **Criteria Name**: Enter a name for the criteria node - **Only for open opportunity**.
    - **Criteria for Executing Actions**: Select the type of criteria that you want to define. You can select **Formula evaluates to true, Conditions are met** (a filter to define the process criteria), or **No criteria-just execute the actions!** In this case, select **Conditions are met.**

- **Set Conditions**: This field lets you specify which combination of the filter conditions must be true for the process to execute the associated actions. In this case, select **[Opportunity].StageName Does not equal Closed Won.**
- **Conditions**: In the **Conditions** section, select **All of the conditions are met (AND)**. This field lets you specify which combination of the filter conditions must be true for the process to execute the associated actions.

It will appear as follows:

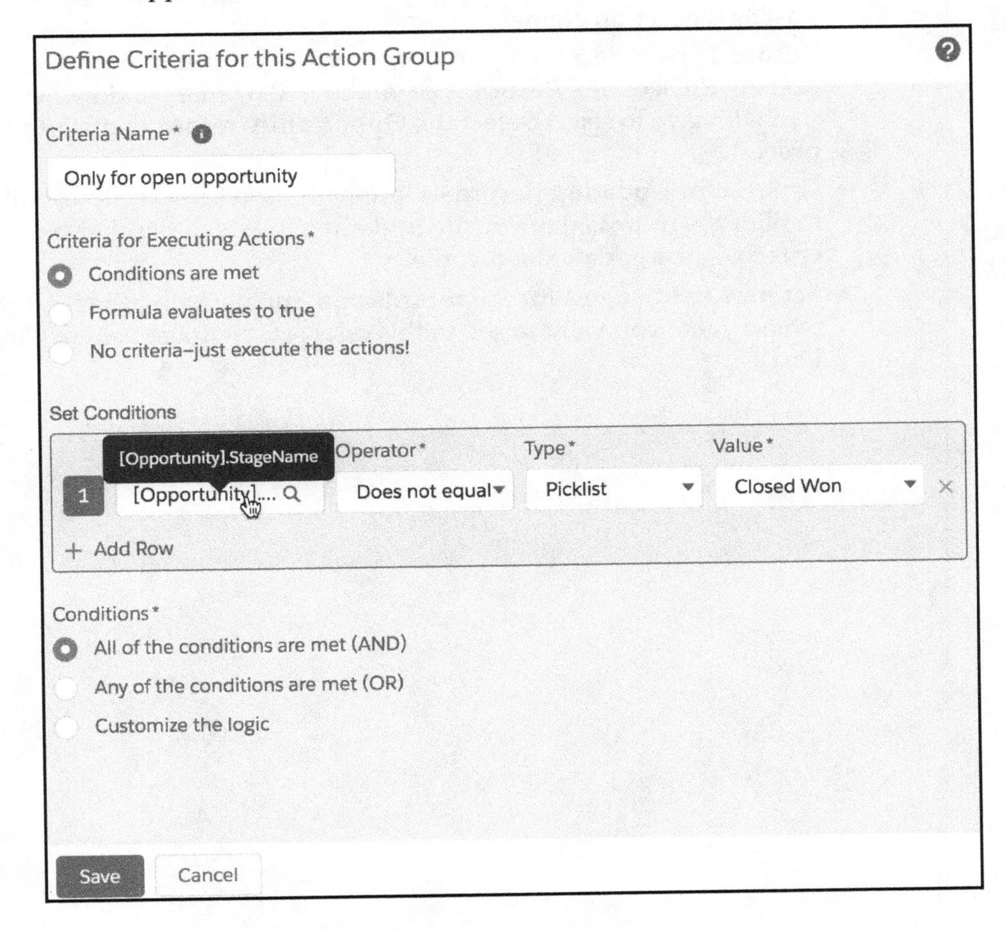

6. Once you are done, click on the **Save** button.
7. Once you are done with the **Process Criteria** node, the next step is to add an **Immediate Action** to update the opportunity stage to Closed Lost. For this, we will use the **Update Records** action in Process Builder. Click on **Add Action** available under **IMMEDIATE ACTIONS**; it will open an additional window on the right-hand side of the process canvas screen, where you will have to enter the following details:
    - **Action Type**: Select the type of action; in this case, select **Update Records**.
    - **Action Name**: Enter a name for this action. Enter **Stage to Closed Lost** as the **Action Name**.
    - **Record Type**: Select the record (or records) that you need to update. Click on the **Record type**, and it will open a window, where you will have to select **Select the Opportunity record that started your process.**
    - **Criteria for Updating Records**: Optionally, you can specify conditions to filter the records that you are updating. In this case, select **No criteria—just update the records!**
    - **Set new field values for the records you update**: Select the field whose value you want to set. In this case, set the **Stage** field to **Closed Lost**.

The preceding steps will look as follows:

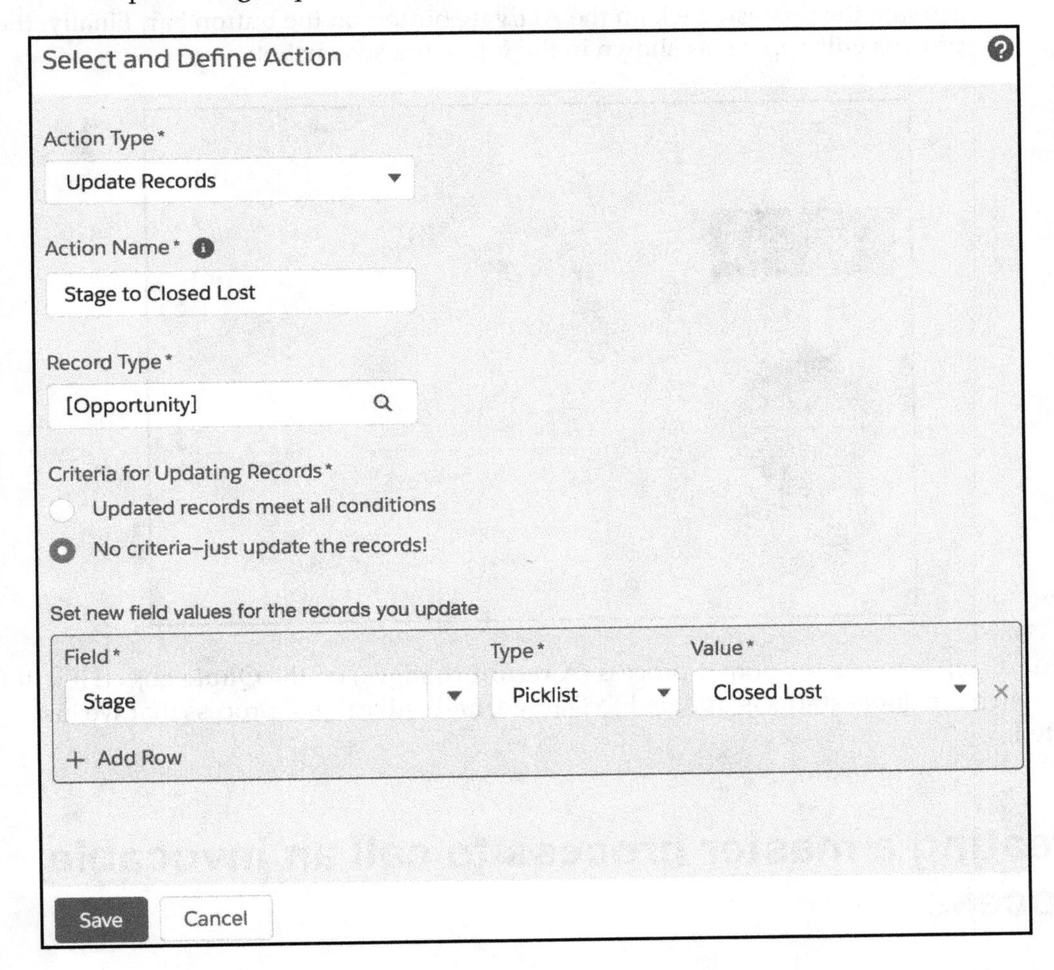

8. Once you are done with the process creation, the final step is to activate it. To activate the process, click on the **Activate** button on the button bar. Finally, the process will appear, as shown in the following screenshot:

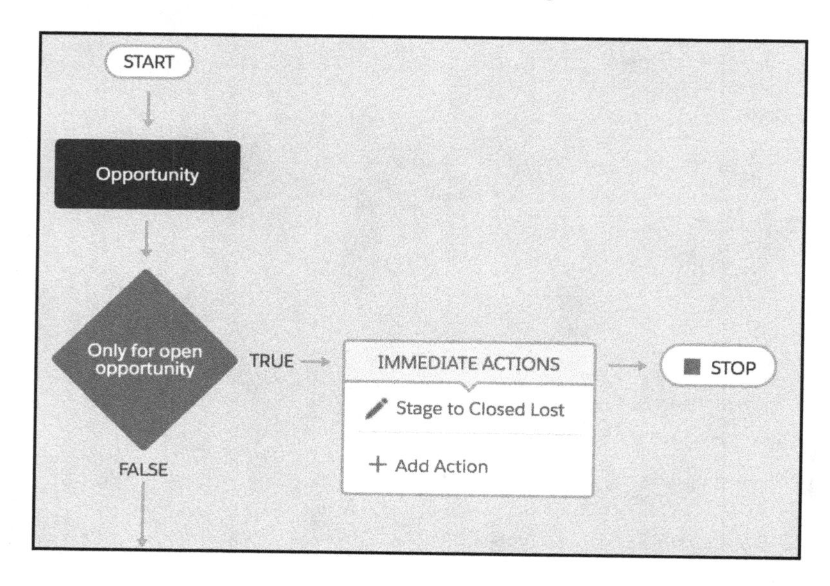

The next step is to create another process (**A record changes**) on the **Quote** object; it will fire only when the quote **status** is updated to **Denied,** and will call the process that we just created.

Creating a master process to call an invocable process

Perform the following steps to create a master process to call the process we just created:

1. Navigate to **Setup** (gear icon) | **Setup** | **PLATFORM TOOLS** | **Process Automation** | **Process Builder**, click on the **New** button, and enter the following details:
    - **Name**: Enter the name of the process - **Master process to call invocable process**.
    - **API Name**: This will be auto-populated, based on the name.

- **Description**: Write some meaningful text, so that other developers or administrators can easily understand why this process was created.
- **This process starts when**: Configure the process to start when a record is created or edited. In this case, select **A record changes.**

2. Once you are done, click on the **Save** button; it will redirect you to the process canvas that will allow you to create the process.
3. After **Define Process Properties**, the next task is to select the object upon which you want to create a process and define the **Evaluation Criteria**. For this, click on the **Add Object** node. It will open an additional window on the right-hand side of the process canvas screen, where you have to enter the following details:
    - **Object**: Start typing, and then select the **Quote** object.
    - **Start the process**: For **Start the process**, select **when a record is created or edited**. This means that the process will fire whenever a record gets created or edited.
    - **Recursion - Allow process to evaluate a record multiple times in a single transaction?**: Select this checkbox only when you want the process to evaluate the same record up to five times in a single transaction. In this case, leave the box unchecked.
4. Once you are done, click on the **Save** button.
5. After defining the **Evaluation Criteria**, the next step is to add the **Process Criteria**. To define the process criteria, click on the **Add Criteria** node; it will open an additional window on the right-hand side of the process canvas screen, where you will have to enter the following details:
    - **Criteria Name**: Enter a name for the criteria node - **Only for denied quotes**.
    - **Criteria for Executing Actions**: Select the type of criteria that you want to define. You can select either **Formula evaluates to true**, **Conditions are met** (a filter to define the process criteria), or **No criteria-just execute the actions!** In this case, select **Conditions are met.**
    - **Set Conditions**: This field lets you specify which combination of the filter conditions must be true for the process to execute the associated actions. In this case, select **[Quotes].Status** equals **Denied.**
    - **Conditions**: In the **Conditions** section, select **All of the conditions are met (AND).** This field lets you specify which combination of the filter conditions must be true for the process to execute the associated actions.
    - Under **Advanced**, select **Yes,** to execute the actions only when the specified changes are made.

The preceding steps will look as follows:

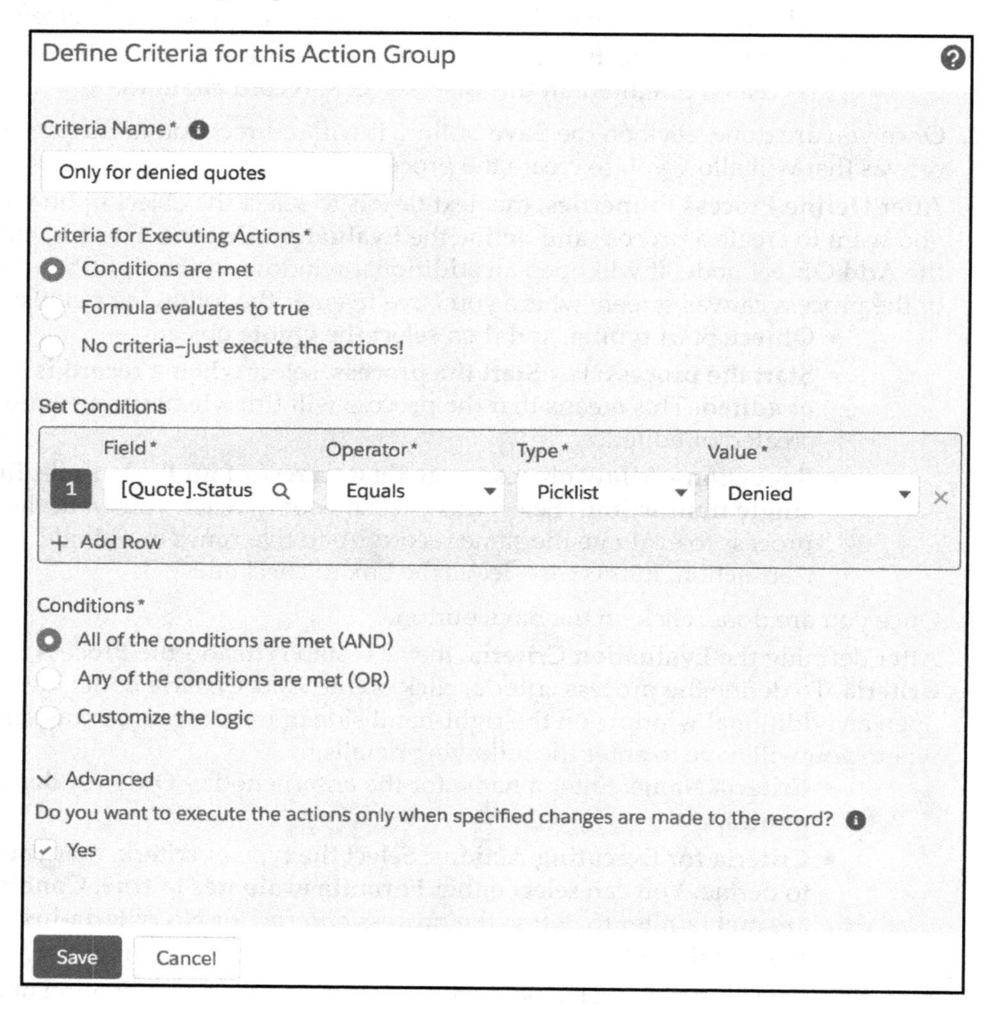

6. Once you are done, click on the **Save** button.
7. The next step is to add an immediate action to update opportunities to Closed Lost. For this, we will use the **Processes** action, available in Process Builder. Click on **Add Action,** available under **IMMEDIATE ACTIONS**; it will open an additional window on the right-hand side of process canvas screen, where you will have to enter the following details:
    - **Action Type**: Select the type of action; in this case, select **Processes**.

- **Action Name**: Enter a name for this action. Enter **Launch opportunity invocable process** as the **Action Name**.
- **Process**: Select the process that you want to execute. In this case, select **Update Opportunity Stage**. You can only select active invocable processes.
- **Set Process Variables**: Select your **Process Variable**. In this case, map the `SObject` variable with **Opportunity ID.**

It will look like the following screenshot:

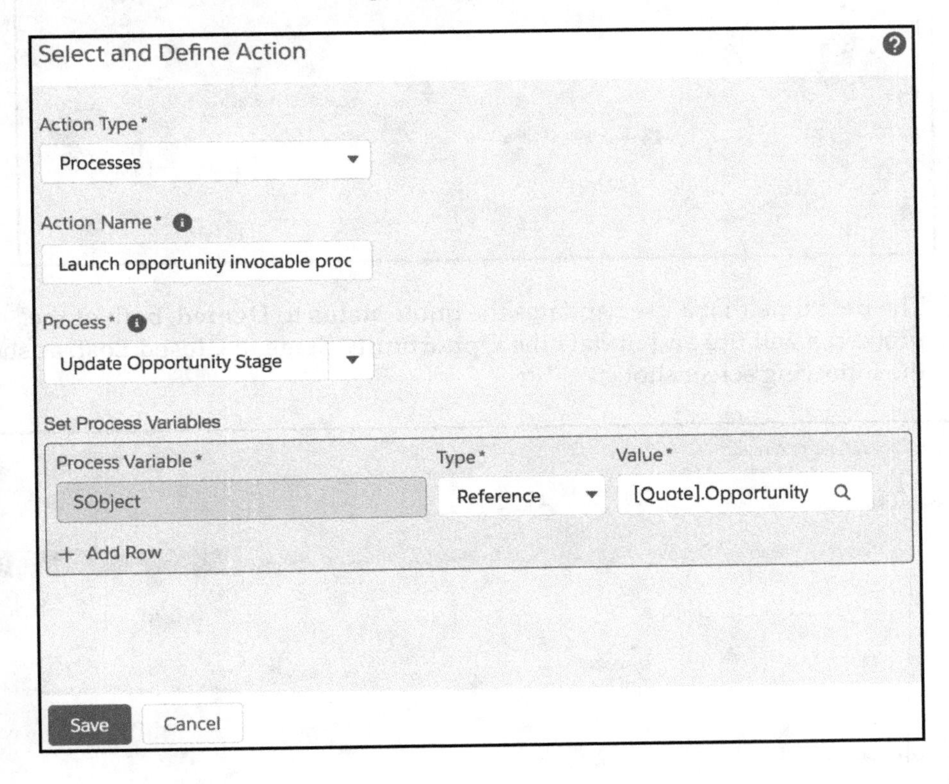

8. Once you are done with the process creation, the final step is to activate it. Click on the **Activate** button on the button bar. Finally, the process will appear, as shown in the following screenshot:

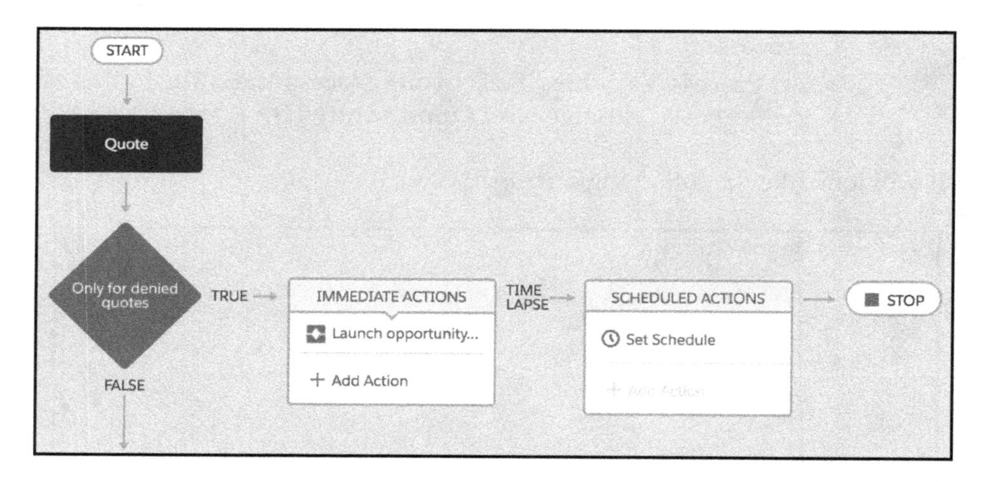

The next time that a user updates the quote **status** to **Denied**, both of the processes will fire and update the **Opportunity Stage** to **Closed Lost**, as shown in the following screenshot:

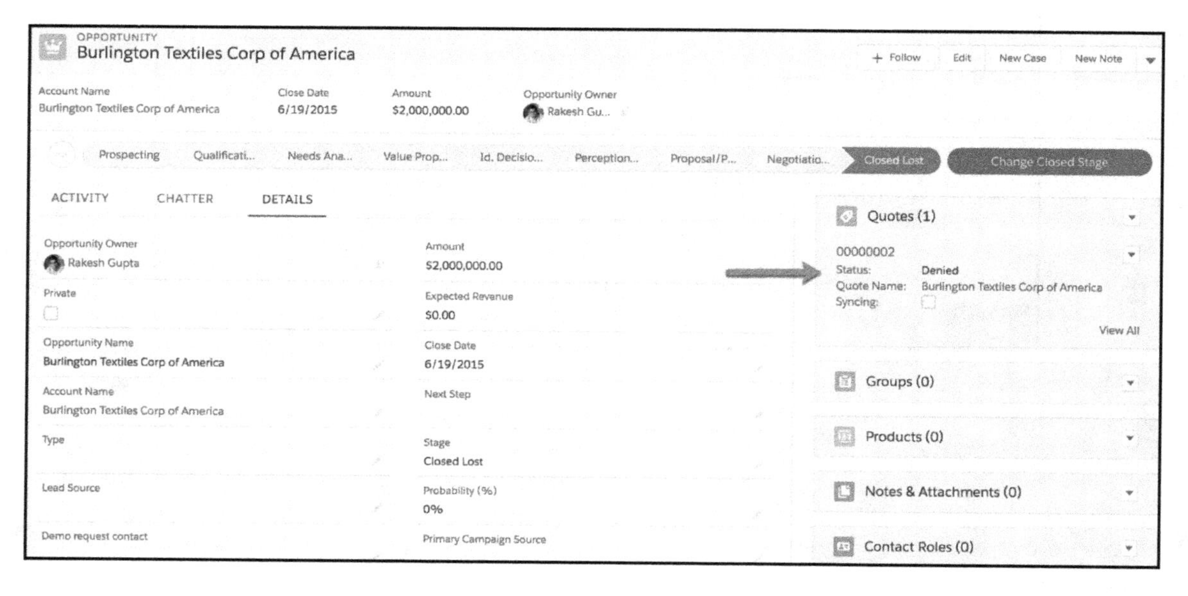

Before testing the process, make sure that you have activated it.

Hands on 10 – using custom metadata types in a Flow

Custom metadata types are similar to custom objects. They allow application developers to create custom sets of data, as well as to create and associate custom data with an organization. All custom metadata type data is available in the application cache, which allows for efficient access, without the cost of repeated queries to the database. It is mainly used to store information that will be frequently accessed from Apex code. It will perform better than a custom object, as it doesn't have to be queried. Building a custom metadata type is very similar to building a custom object. The main difference that you will notice is the `__mdt` suffix at the end of the custom metadata type, as opposed to the usual `__c`, for custom objects. As of the Summer 2017 release, custom metadata is not yet available in Process Builder, but it is available in Flow.

Let's look at a business scenario. Suppose that Helina Jolly is working as a system administrator at Universal Containers. She has received a requirement to auto-create a case whenever a new Chatter post (not a comment) contains a banned word. She has also received a list of banned words (almost 50 words) from her manager. If Chatter posts contain numerous banned words, the process should create a single case.

For the preceding business scenario, you can use all 50 words in a Process Builder criteria, and use the **Create a Record** action to auto-create a case. But the problem with this approach is, if, after a few days, the business wants to add/remove a few banned words, then you will have to start from scratch. Custom metadata provides the flexibility to add/remove banned words, without having to change your process definition or Flow.

To learn more about custom metadata, go to `https://developer.salesforce.com/docs/atlas.en-us.api_meta.meta/api_meta/meta_custommetadatatypes.htm`.

To solve the preceding business requirement, we will use custom metadata types to store the banned words, Flow to compare the banned words with the Chatter post body, and Process Builder to launch the Flow. Perform the following steps to solve the preceding business requirement:

1. To create new custom metadata type, navigate to **Setup** (gear icon) | **Setup** | **PLATFORM TOOLS** | **Custom Code** | **Custom Metadata Types**, and click on the **New Custom Metadata Type** button; it will redirect you to a new window, where you will have to enter the following details:
    - **Label**: Enter a label for the custom settings. Enter **Chatter Post** as the criteria name.
    - **Plural Label**: Enter a plural label. If you create a tab for the custom metadata type, this name will be used. Enter **Chatter Posts** as the plural name.
    - **Object Name**: Enter the unique object name; it will be used when the custom metadata type is referenced by formula fields, validation rules, Apex, or the SOAP API. This will be auto-populated, based on the **Label**.
    - **Description**: Write some meaningful text, so that another developer or administrator can easily understand why this custom metadata type was created.
    - **Visibility**: For **Visibility**, select **All Apex code and APIs can use the type, and it is visible in Setup**.
2. Once you are done, click on the **Save** button.
3. Create a **Text** field, **Banned Word**, to store the words that are not allowed in the Chatter post, and make this field a required field. At the end, the **Chatter Post** custom metadata type should look as follows:

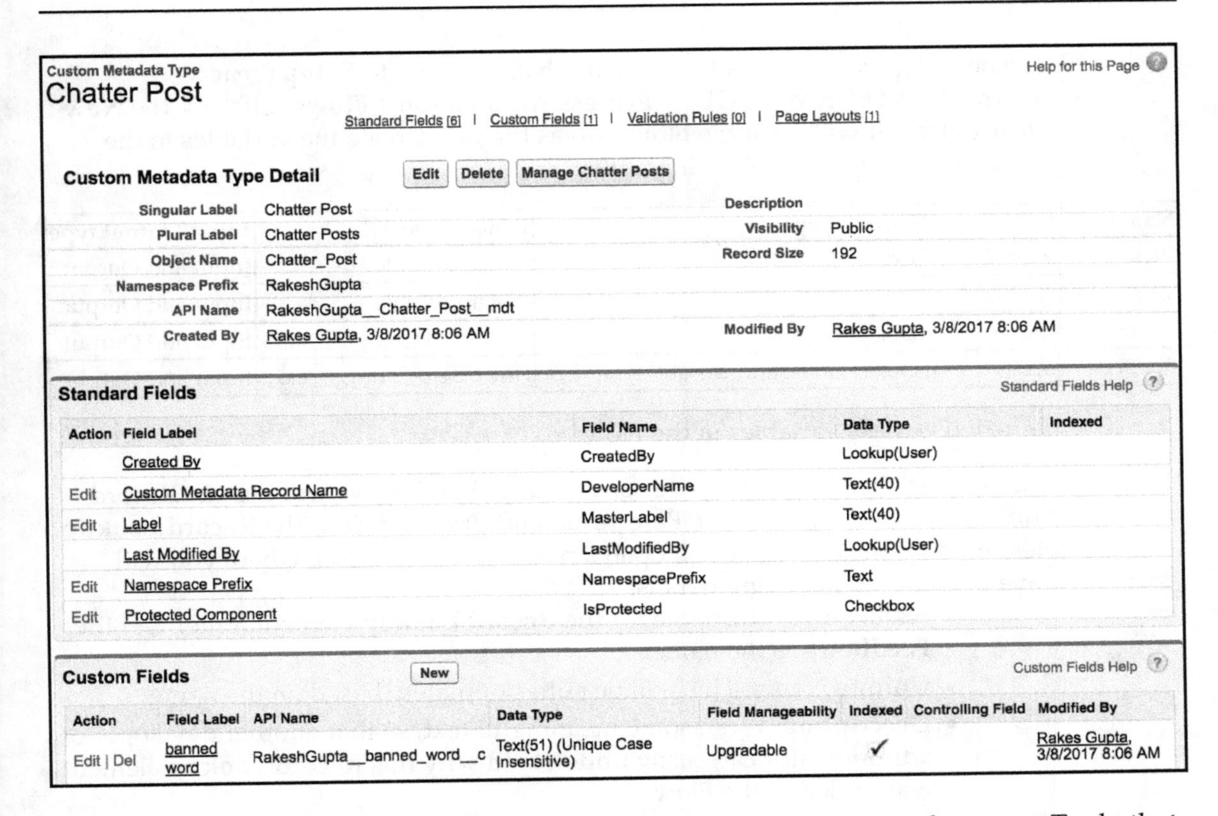

4. The next step is to insert a few records into the custom metadata type. To do that, click on the **Manage Chatter Posts** button on the custom metadata type detail page, and then click on **New,** to create some custom setting records, as shown in the following screenshot:

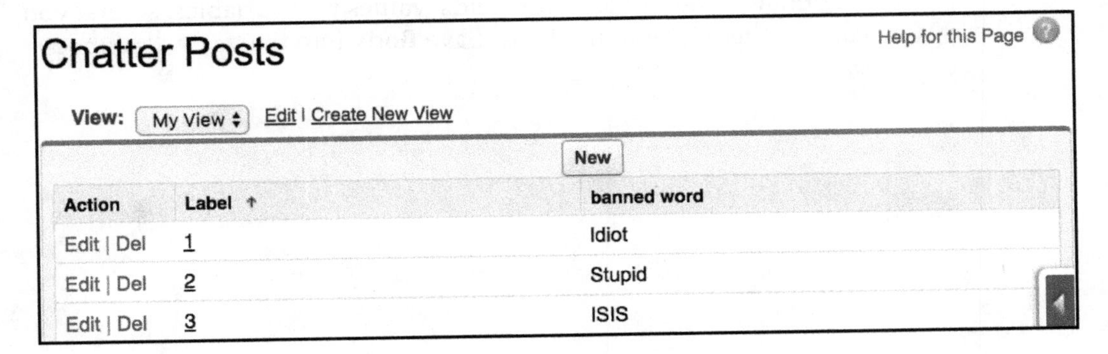

5. The next step is to create a Flow. To do that, navigate to **Setup** (gear icon) | **Setup** | **PLATFORM TOOLS** | **Process Automation** | **Flows**. Click on the **New Flow** button; it will open the Flow canvas for you. Create the variables in the Flow, as shown in the following table:

Name	**Variable type**	**Object type**	**Input/Output type**
`VarT_Feedbody`	Text	Not applicable	Input and Output
`VarT_FeedItemId`	Text	Not applicable	Input and Output
`SovBannedWord`	SObject variable	`Chatter_Post__mdt`	Input and Output
`SOCVBannedWords`	SObject Collection variable	`Chatter_Post__mdt`	Input and Output

We will use these variables in the Flow.

6. The next task is to get the Chatter post bodies. For this, we will use the **Record Lookup** element. Click on the **Palette** tab and drag and drop the **Record Lookup** element onto the canvas; it will open a new window for you, where you will have to enter the following details:
    - **Name**: Enter the name for the **Record Lookup** element. Enter **Get the Feedbody** as the name.
    - **Unique Name**: This will be auto-populated, based on the name.
    - **Description**: Write some meaningful text, so that another developer or administrator can easily understand why this **Record Lookup** element was added to the Flow.
    - **Look up**: Select the object for which you want to search the record. In this case, select the **FeedItem** object. The next task is to define the search criteria. For this, using **Id** equal `VarT_FeedItemId`.
    - **Assign the record's fields to variables to reference them in your flow**: Optionally, you can save the fields' values into variables, so that you can use them later in the Flow. Save **Body** into the `VarT_Feedbody` variable.

To map the fields, you can use the information in the following screenshot:

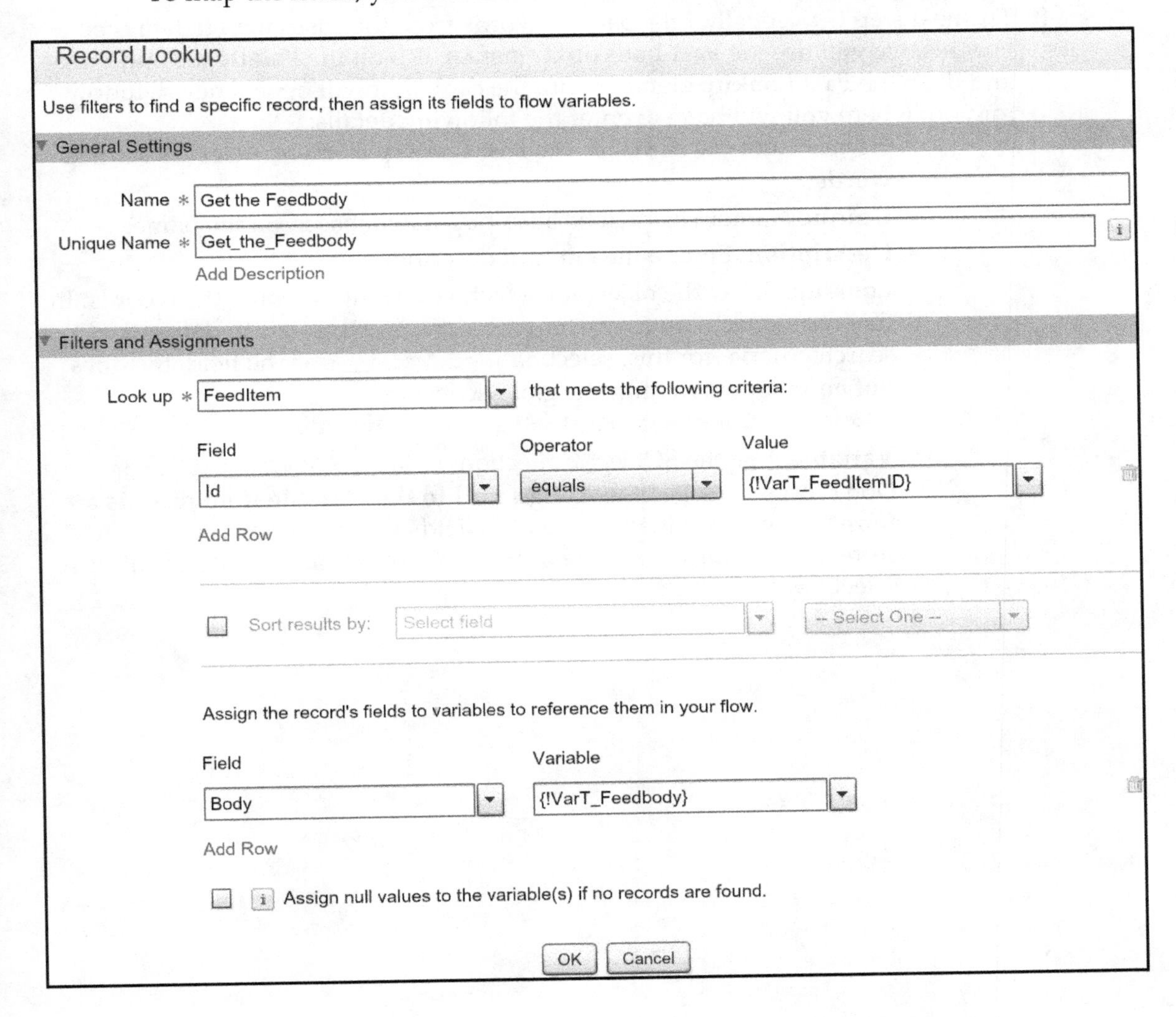

7. Once you are done, click on the **OK** button.
8. The next step is to get all of the banned words from the custom metadata type. For this, we will use the **Fast Lookup** element. Click on the **Palette** tab and drag and drop the **Fast Lookup** element onto the canvas; it will open a new window for you, where you will have to enter the following details:
    - **Name**: Enter the name for the **Fast Lookup** element – **Get all banned words**.
    - **Unique Name**: This will be auto-populated, based on the name.
    - **Description**: Enter a meaningful description.
    - **Look up**: Select the object for which you want to search the records. In this case, select `Chatter_Post_mdt`. The next task is to define the search criteria; for this, select `banned_word__c` as the field, the **does not equal** operator, and the global constant `{!$GlobalConstant.EmptyString}` as the value.
    - **Variable**: Use the SObject Collection variable, `SOCVBannedWords`. Don't forget to select the **Assign null to the variable if no records are found** checkbox. Finally, select the fields whose values you want to store in the SObject variable or SObject Collection variable; in this case, select `banned_word__c`.

To map the fields, you can use the information in the following screenshot:

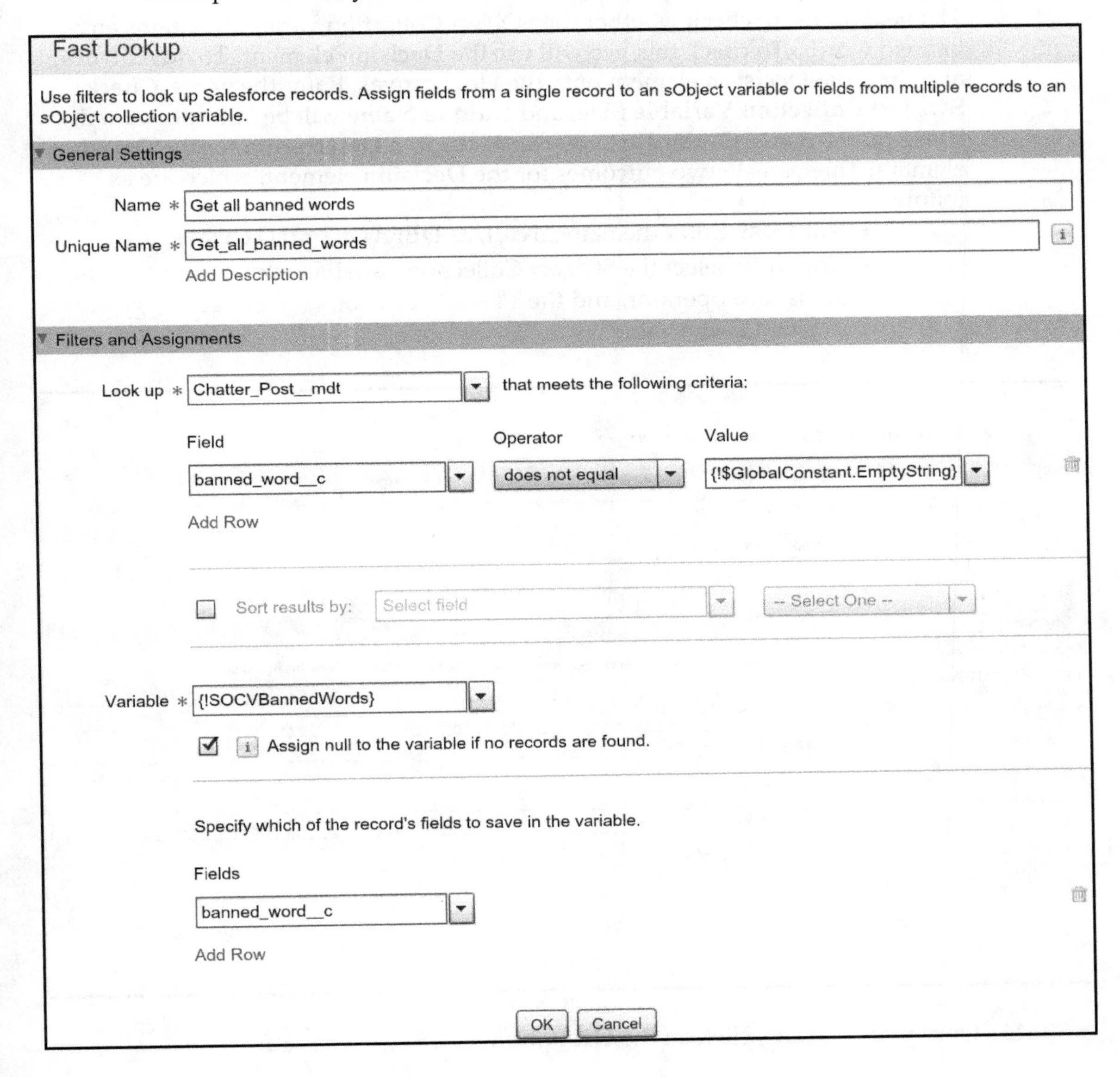

9. Once you are done, click on the **OK** button.
10. The next task is to check whether the SObject Collection variable contains any banned words. To check this, we will use the **Decision** element. To do that, drag and drop the **Decision** element onto the Flow canvas. Enter the name, **Check SObject Collection Variable Size**, and **Unique Name** will be auto-populated, based on the name. Optionally, you can also add a **Description** for the **Decision** element. Then, create two outcomes for the **Decision** element, which are as follows:
    - **Not Exist**: Enter the name, **Null,** as **DEFAULT OUTCOME**.
    - **Not Null**: Select the SObject Collection variable, `SOCVBannedWords`, the **is null** operator, and the `{!$ GlobalConstant.False}` global constant as the value.:

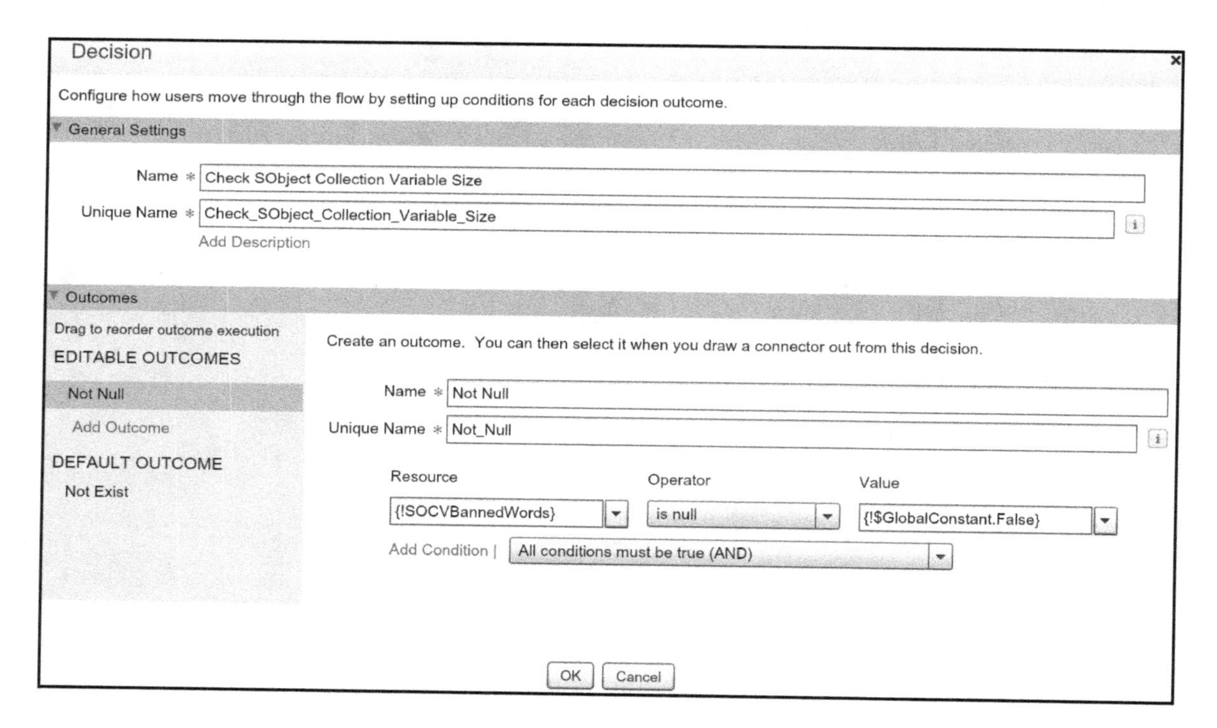

11. Once you are done, click on the **OK** button.

12. If the SObject Collection variable is not null, that means that it contains some banned words. We will use the **Loop** element to extract records from the SObject Collection variable (`SOCVBannedWords`) and store it to the SObject variable (`SovBannedWord`). Click on the **Palette** tab and drag and drop the **Loop** element onto the Flow canvas. It will open a new window for you, where you will have to enter the following details:
    - **Name**: Enter the name for the **Loop** element. In this case, enter **Loop over collection** as the name.
    - **Unique Name**: This will be auto-populated, based on the name.
    - **Description**: Write some meaningful text, so that another developer or administrator can easily understand why this **Loop** element was created.
    - **Loop through**: Select the SObject Collection variable, `SOCVBannedWords`. Select the order as **Ascending,** to loop through the collection.
    - **Loop Variable**: Select the SObject variable or variable as **Loop Variable**. In this case, select the SObject variable, `SovBannedWord`, as the loop variable.

To map the variable, you can use the information in the following screenshot:

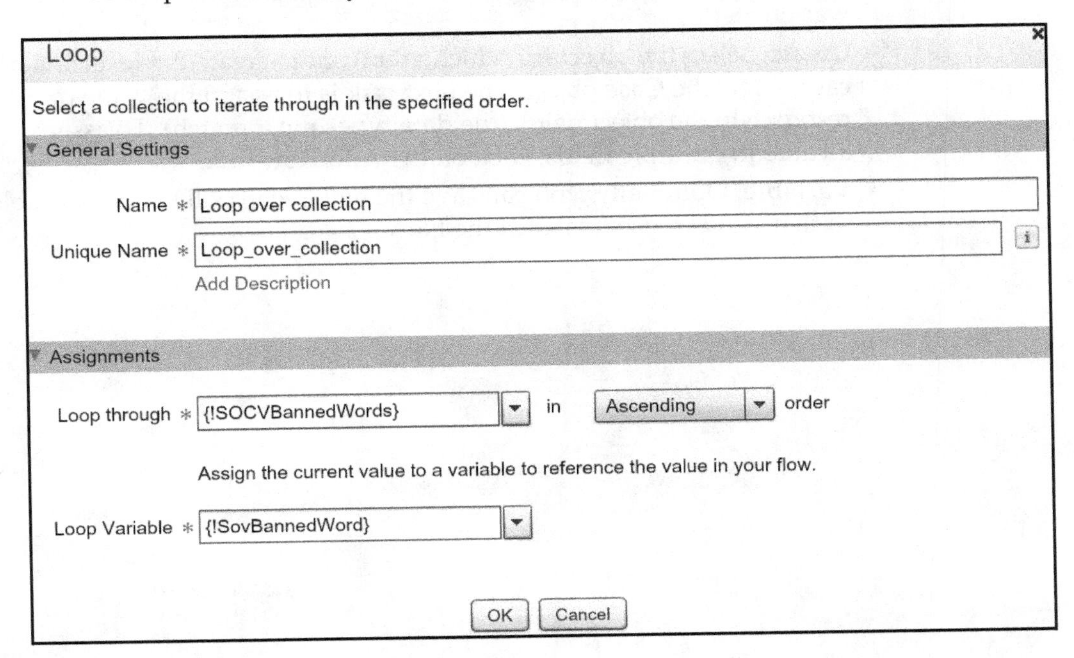

13. Once you are done, click on the **OK** button.
14. The next task is to check whether `VarT_feedbody` contains any banned words. To check this, we will use the **Decision** element. To do that, drag and drop the **Decision** element onto the Flow canvas. Enter the name, **Lookup for banned words**, and **Unique Name** will be auto-populated, based on the name. Optionally, you can also add a description for the **Decision** element. Then, create two outcomes for the **Decision** element, which are as follows:
    - **Not Found**: Enter the name, **Null**, as **DEFAULT OUTCOME.**
    - **Found**: Select the variable `VarT_Feedbody`, which contains the operator, and `{!SovBannedWord.banned_word__c}` as the value.
15. Once you are done, click on the **OK** button.
16. If the `VarT_Feedbody` variable contains a banned word, then we will have to create a new case. Click on the **Palette** tab and drag and drop the **Record Create** element onto the canvas; it will open a new window for you, where you will have to enter the following details:
    - **Name**: Enter the name for the **Record Create** element. In this case, enter **Create new case** as the name.
    - **Unique Name**: This will be auto-populated, based on the name.
    - **Description**: Write some meaningful text, so that another developer or administrator can easily understand why this **Record Create** element was created.
    - **Create**: Select the object for which you want to create the record. In this case, select the **Case** object. The next task is to assign the value or resource to the object fields (the data types must match). To assign a value to multiple fields, click on the **Add Row** link.
    - **Variable**: Optionally, you can save the new record's ID into a variable, so that you can use it later in the Flow.

To map the fields, you can use the information in the following screenshot:

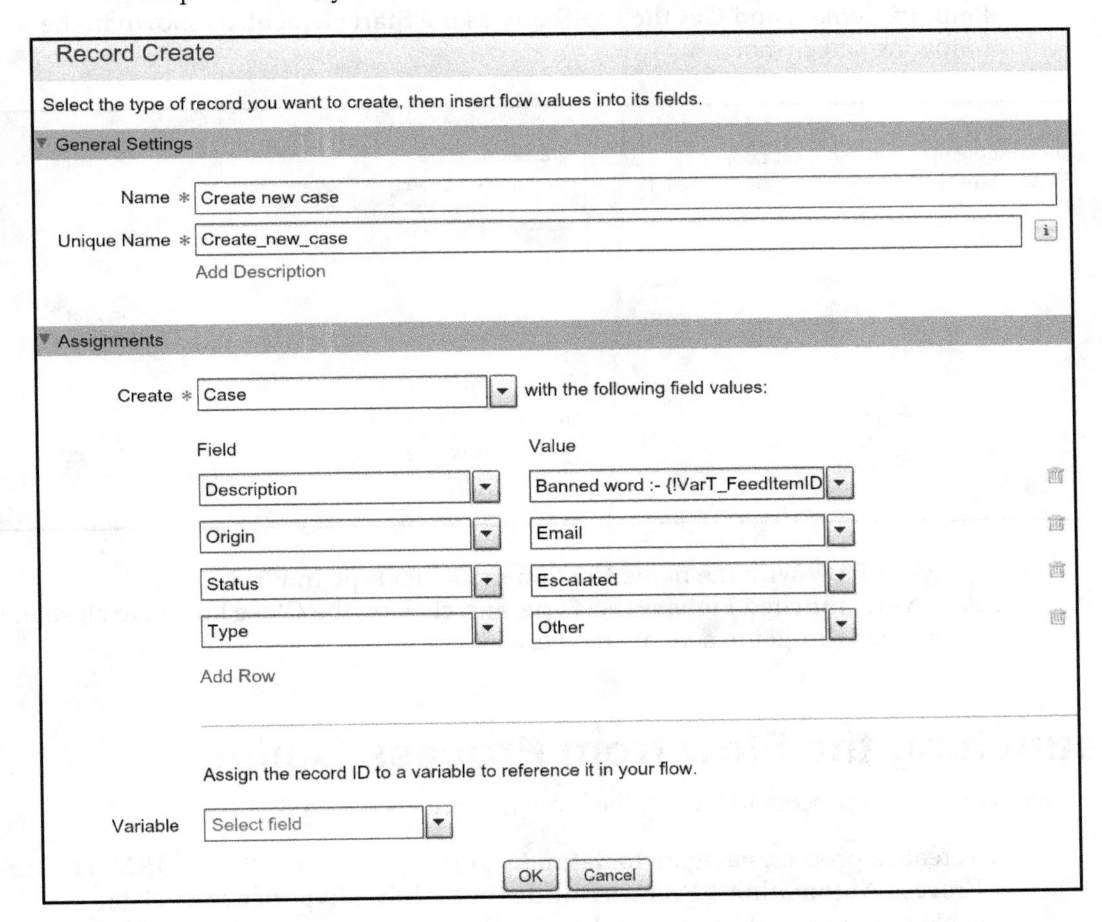

17. Once you are done, click on the **OK** button.

18. Use the connector to connect the elements used in the Flow. Set the **Record Lookup** element and **Get the Feedbody** as the **Start** element, as shown in the following screenshot:

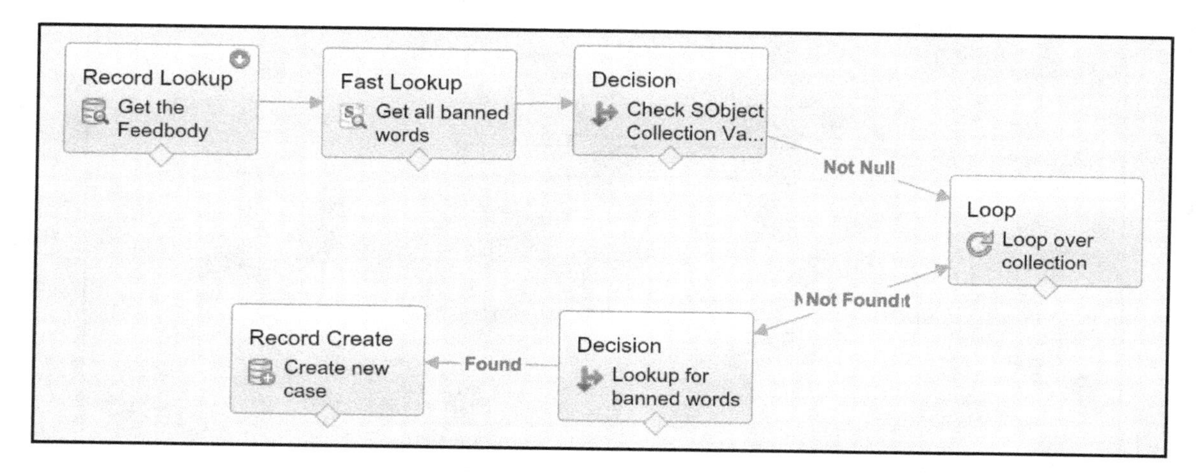

19. Save your Flow with the name **Custom metadata type in Flow**, select **Autolaunched Flow** as the **Type**, and click on the **Close** button to close the canvas. Don't forget to activate the Flow.

Launching the Flow from Process Builder

Now, we will create a process to launch the Flow that we just created:

1. To create a process, navigate to **Setup** (gear icon) | **Setup** | **PLATFORM TOOLS** | **Process Automation** | **Process Builder** in **Lighting Experience**, click on the **New** button, and enter the following details:
   - **Name**: Enter the name of the process - **Custom metadata type in Flow -PB**.
   - **API Name**: This will be auto-populated, based on the name.
   - **Description**: Write some meaningful text, so that other developers or administrators can easily understand why this process was created.
   - **This process starts when**: Configure the process to start when a record is created or edited. In this case, select **A record changes**.
2. Once you are done, click on the **Save** button; it will redirect you to the process canvas that will allow you to create the process.

3. After **Define Process Properties**, the next task is to select the object upon which you want to create a process and define **evaluation criteria**. For this, click on the **Add Object** node. It will open an additional window on the right-hand side of the process canvas screen, where you will have to enter the following details:
    - **Object**: Start typing, and then select the **FeedItem** object.
    - **Start the process**: For **Start the process**, select **only when a record is created**. This means that the process will only fire at the time of record creation.
    - **Recursion - Allow process to evaluate a record multiple times in a single transaction?**: Select this checkbox only when you want the process to evaluate the same record up to five times in a single transaction. In this case, leave this box unchecked.

   The preceding steps will look as follows:

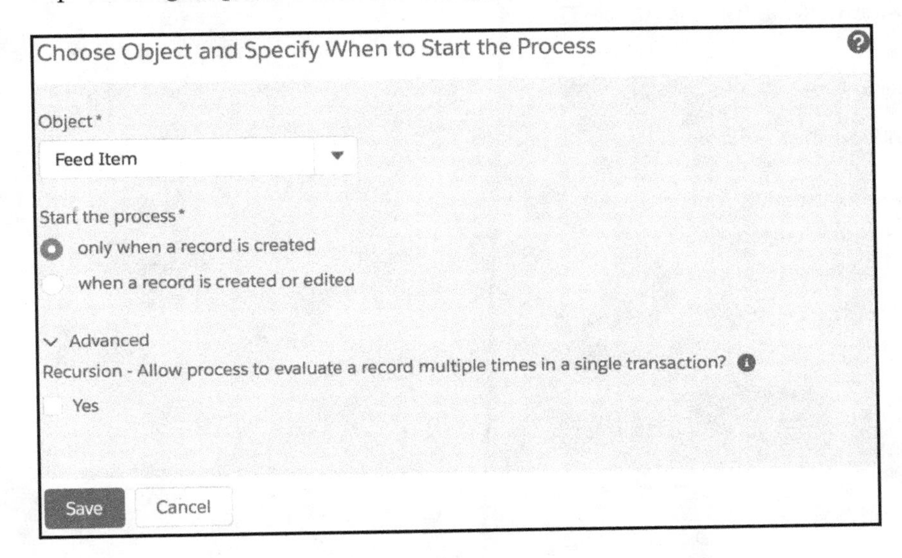

4. Once you are done, click on the **Save** button.
5. After defining the **evaluation criteria**, the next step is to add the **process criteria**. Once the process criteria are true, the process will execute the associated actions. To define the process criteria, click on the **Add Criteria** node. It will open an additional window on the right-hand side of the process canvas screen, where you will have to enter the following details:
    - **Criteria Name**: Enter a name for the criteria node. Enter `Type equals to TextPost` as the criteria name.

- **Criteria for Executing Actions**: Select the type of criteria that you want to define. You can select either **Formula evaluates to true, Conditions are met** (a filter to define the process criteria), or **No criteria-just execute the actions**. In this case, select **Conditions are met.**
- **Set Conditions**: This field lets you specify which combination of the filter conditions must be true for the process to execute the associated actions. In this case, set `[FeedItem].Type` to `Text Post`.
- **Conditions**: In the **Conditions** section, select **All of the conditions are met (AND)**. This field lets you specify which combination of the filter conditions must be true for the process to execute the associated actions.

The preceding steps will look as follows:

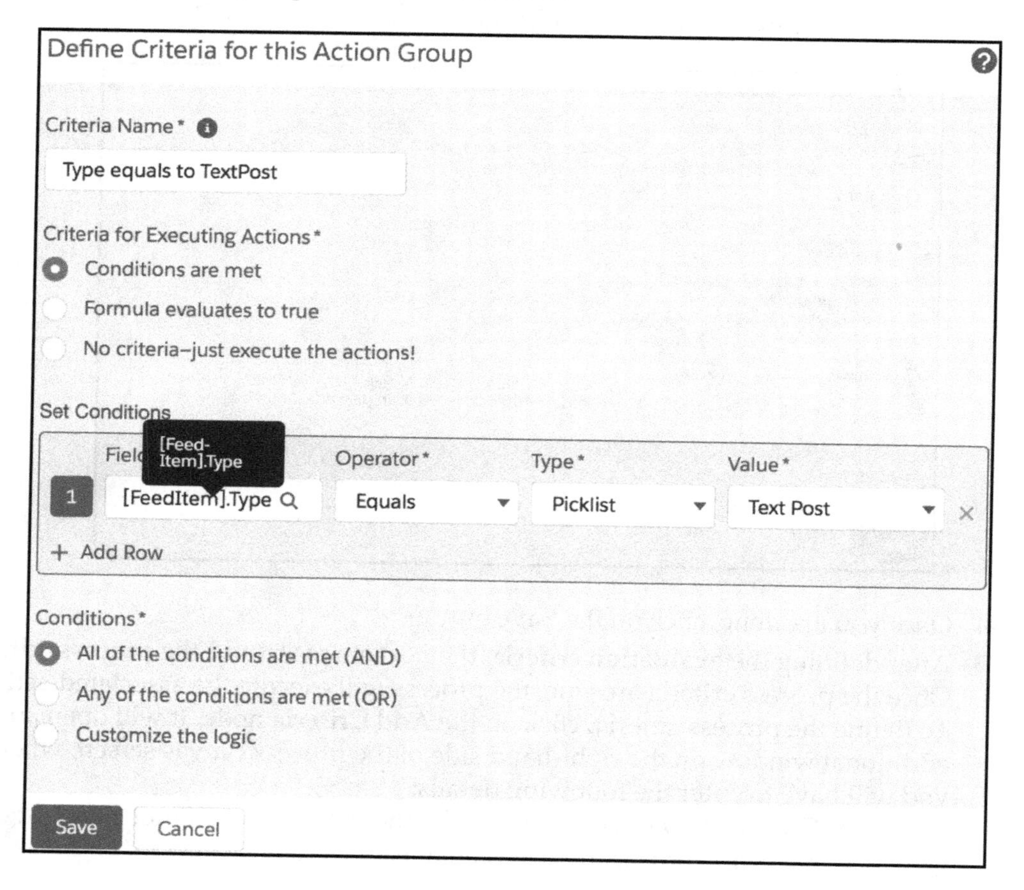

6. Once you are done, click on the **Save** button.
7. Once you are done with the **process criteria** node, the next step is to add an immediate action to launch a Flow, so that it will pass the Feeditem ID to the Flow whenever any text post gets created. For this, we will use the **Flows** action available in Process Builder. Click on **Add Action,** available under **IMMEDIATE ACTIONS**; it will open an additional window on the right-hand side of the process canvas window, where you will have to enter the following details:
    - **Action Type**: Select the type of action; in this case, select **Flows.**
    - **Action Name**: Enter a name for this action. Enter **Launch a Flow** as the action name.
    - **Flow**: Select the Flow that you want to execute; in this case, select the Flow, **Custom metadata type in Flow**.
    - **Set Flow Variables**: Use this to pass the value in your Flow variables. For the current use case, map the variable `VarT_FeedItemId` with **[FeedItem].Id**.

The preceding steps will look as follows:

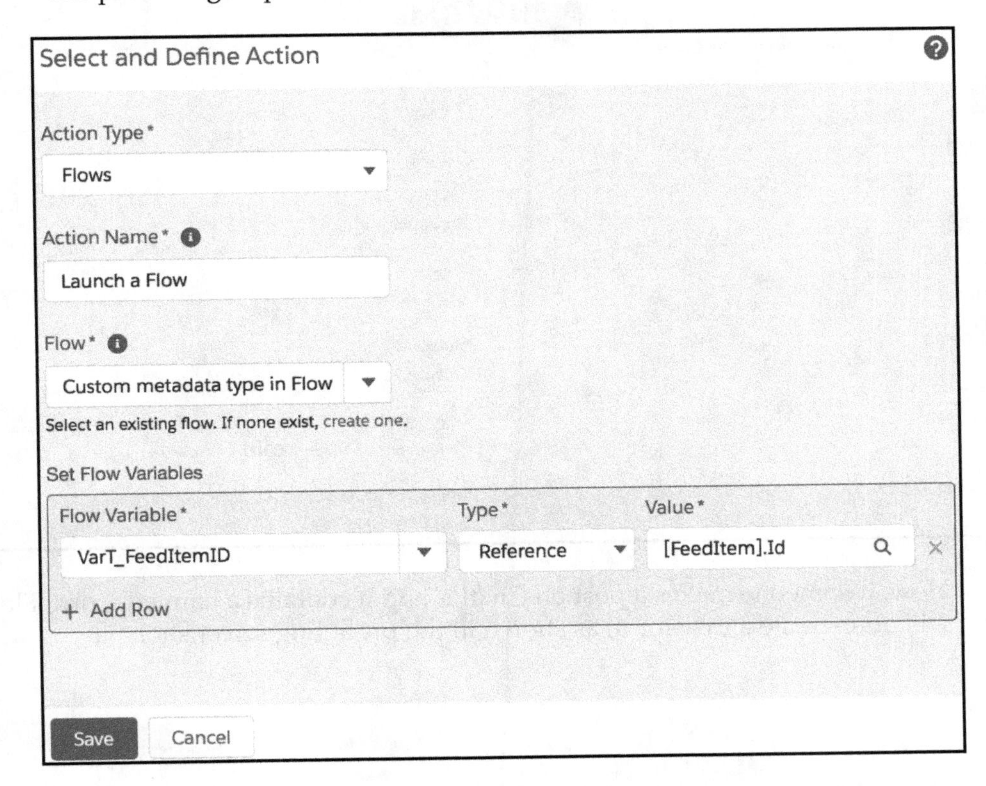

To assign values to multiple variables, click on the **Add Row** link.

8. Once you are done with the process creation, the final step is to activate it. Click on the **Activate** button, available on the button bar:

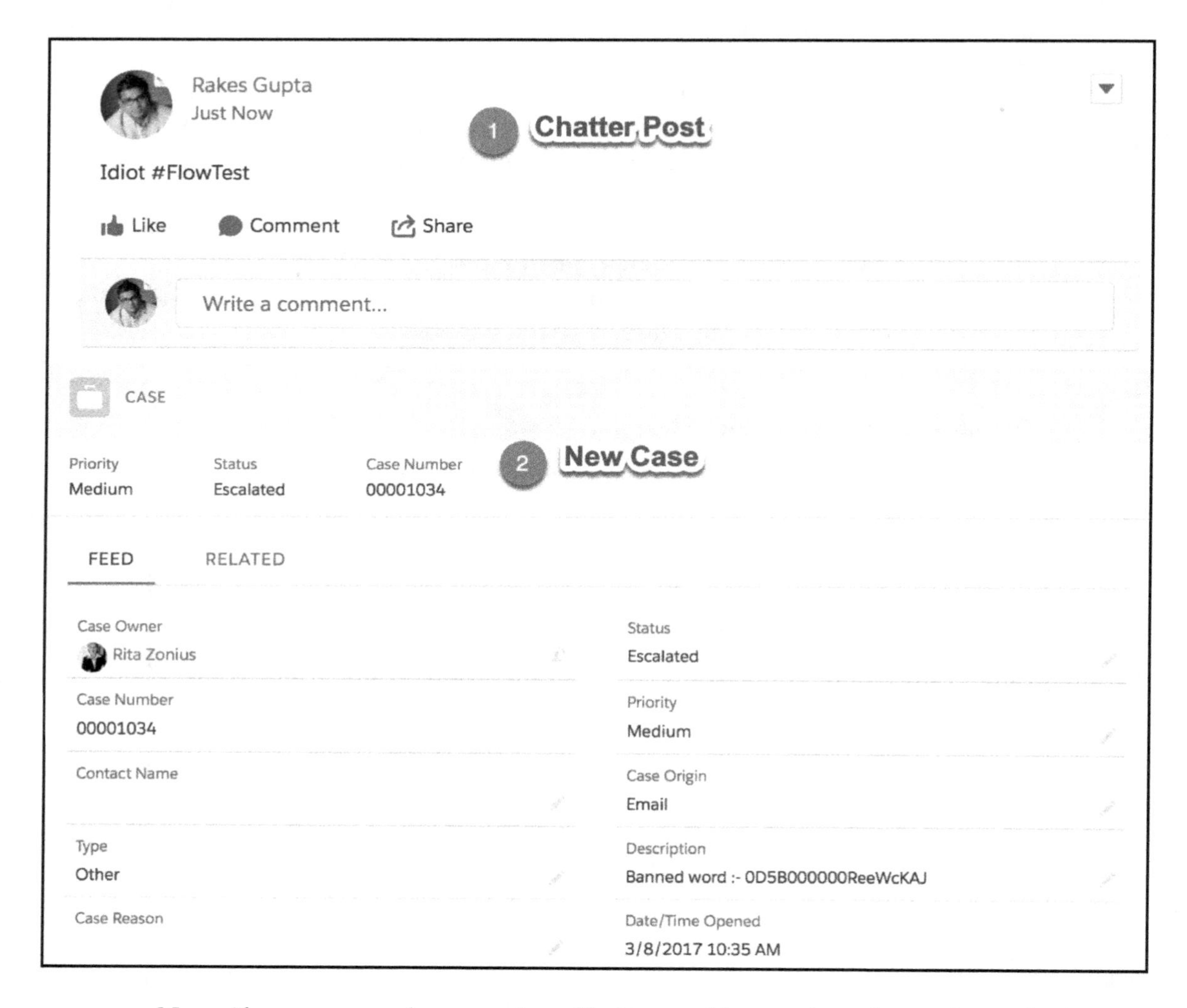

Now, if someone makes a post on Chatter and it contains a banned word, Flow will auto-create a case for it, as shown in the preceding screenshot.

A few points to remember

The following are some noteworthy points to remember:

- You can't evaluate the next criteria if a criteria group contains scheduled actions.
- A DML operation on a setup object is not permitted after you have updated a non-setup object (or vice versa). If you want to do that, then use a time-dependent action. You can find the list of setup objects in Salesforce at `https://developer.salesforce.com/docs/atlas.en-us.api_tooling.meta/api_tooling/reference_objects_setup.htm`.
- Don't perform the DML operation inside the **Loop** element. It will easily hit the governor limit - that is, **System.LimitException: Too many SOQL queries: 101**.
- You can only launch Autolaunched Flow from the Process Builder.
- A Flow runs in user mode, and Process Builder runs in the system mode. Let's look at an example if you are trying to update Opportunity the next step:
  - **If you use Process Builder**: If the running user doesn't have access to the next step field, Process Builder will be able to update it.
  - **If you use Flow**: If the running user doesn't have access to the next step field, they will get an error.
- If the Flow doesn't have a Start element, you won't get a link to activate the Flow.
- Activating this process automatically deactivates any other active version. The deactivated version will be available in your version history.
- Actions are executed in the order in which they appear in Process Builder.
- Only active invocable processes are available to select under **action processes**.
- Avoid creating an infinite loop when allowing your process to reevaluate records multiple times in a transaction. For example, if your process checks whether an opportunity description changes, and then updates an opportunity description and creates a Chatter post every time an opportunity record is created or edited, the process will evaluate and trigger actions, resulting in six Chatter posts.
- You can have up to 200 criteria nodes in a process.
- After you deactivate a process, any scheduled actions will still be there in a queue for execution.
- An organization can have up to 30,000 pending schedules and waiting Flow interviews at one time.
- If a user deletes the record or the object that the schedule is associated with, the schedule will never be processed.

- The formula returns true or false. If the formula returns true, the associated actions will be executed.
- The DML operation on a setup object is not permitted at the same time as when you update a non-setup object (or vice versa). If you want to do that, then use a time-dependent action.
- Processes on a task or the event object support email alerts.
- You can have up to 50 versions of a process, but only one version of a process can be active.

Exercises

1. Create a process that auto-adds new users to the `Sales Best Practices` Chatter group

First, create a Chatter group called `Sales Best Practices`. Use Quick Actions to complete it.

2. Create a process that will automatically delete open opportunities when an account is **out of business**.

Use the Apex class and invoke it from the process.

3. Modify the previous process in such a way that it will not work for the **System Administrator** and **Supply Chain User** profiles.
4. Create a process that will automatically count related contacts in the account and update the value in a **Number of contacts** field. This process will fire whenever the account is updated by all users, except the System Administrator profile.

You may need to use Flow to get the count of contacts in an account.

5. Create a process that will auto-remove all followers from a case when it has successfully closed.
6. Create a process that will do the following:
    1. Send an email to the owner 10, 15, and 20 days before the opportunity close date.
    2. After three reminder emails, if the opportunity is still open, then send an email to the account owner.
    3. Once an opportunity has successfully closed, auto-create the contract from it. Use the following values to create a new contract:
        - **Name**: The same as the account name.
        - **Status**: Active.
        - **Contract Start Date**: Today.
        - **Contract Term (Months)**: 12.
7. Create a process that will do the following:
    1. Update the opportunity stage to Closed Lost if the opportunity is still open 30 days after the close date.
    2. Remove all members from the opportunity team.
    3. Create a case from the lost opportunity to find the cause. Use the following values to create the new contract:
        - **Subject**: The same as the opportunity name.
        - **Status**: New.
        - **Priority**: High.
        - **Case Origin**: Lost opportunity (add one value in the picklist).
        - **Owner**: Assign it to the account owner.
8. Once an order has successfully completed, send an email to the order owner after 10 and 20 days, for payment confirmation.
9. Create a process that will send a private message to new users, with the Chatter best practices link.

You may need to use Flow to send a chat message.

10. Automatically create a Chatter group for each campaign, as soon as the campaign is activated by a user. Set yourself as the Chatter group owner, and the campaign owner as the Chatter group manager. For the Chatter group name, use the campaign name.
11. Create a process that will auto-update account information to **Automation Champion** in an opportunity, if the opportunity gets created with no account.

First, create an account with the name Automation Champion. Use the custom label to store the account ID.

12. Once an opportunity has successfully closed, auto-create assets from opportunity products.

Use Flow to create assets from opportunity products.

13. Create a process that will auto-remove users from public groups, once a user's account gets deactivated by the system administrator.

Use both Flow and Process Builder to remove users.

14. Once an account has updated to **out of business**, update all open opportunities to Closed Lost.

Use the invocable process that you created in the *Creating reusable processes using an invocable process* section.

15. Create a process that auto-converts leads once the **Rating** is updated to **Hot**.
16. Once an opportunity has successfully closed, auto-create an order and copy the opportunity products to order products.

17. If a post is made by the CEO on their Chatter profile, auto-post the same content to the `Sales Executive` Chatter group; set **Created by** to **CEO**.
18. Create an application that auto-adds new users to the `Universal Container Employee` public group.

First, create a public group called Universal Container Employee.

Summary

In this chapter, we went through some advanced concepts of Process Builder, starting with Workbench and Audit Trail, to get more information about Flows and Processes. Then, we moved on and discussed how to define an additional condition using record updates. We discussed how you can execute multiple criteria of a process. We also discussed how you can schedule multiple groups of actions and create reusable processes using an invocable process. Finally, we illustrated a way to use custom metadata types and custom labels in the Flow, and how to create scheduled jobs.

Having completed this chapter, you should have a lot of knowledge about Salesforce Lightning Process Builder. If you want to explore it further, refer to the following blogs:

- `http://www.salesforceweek.ly/`
- `https://automationchampion.com/`
- `http://salesforceyoda.com/`
- `http://jenwlee.wordpress.com`
- `http://www.salesforcesidekick.com`

Don't forget to join the official Salesforce Workflow Automation Chatter group in the Success Community. You can do that at `https://success.salesforce.com/_ui/core/chatter/groups/GroupProfilePage?g=0F9300000001rzc&s=workflo&r=1`.

Other Books You May Enjoy

If you enjoyed this book, you may be interested in these other books by Packt:

Salesforce CRM Admin Cookbook - Second Edition
Paul Goodey

ISBN: 978-1-78862-551-7

- Building home page components and creating custom links to provide additional functionality and improve the Home Tab layout
- Improving the look and feel of Salesforce CRM with the presentation of graphical elements using advanced user interface techniques
- Improving the data quality in Salesforce CRM and automatic data capture
- Implement an approval process to control the way approvals are managed for records in Salesforce CRM
- Increase productivity using tools and features to provide advanced administration
- Extend Lightning Experience Record Pages to tailor user interaction experience
- Create Lightning component to implement Search before Create for customer/person accounts

Salesforce Lightning Reporting and Dashboards
Johan Yu

ISBN: 978-1-78829-738-7

- Navigate in Salesforce.com within the Lightning Experience user interface
- Secure and share your reports and dashboards with other users
- Create, manage, and maintain reports using Report Builder
- Learn how the report type can affect the report generated
- Explore the report and dashboard folder and the sharing model
- Create reports with multiple formats and custom report types
- Explore various dashboard features in Lightning Experience
- Use Salesforce1, including accessing reports and dashboards

Leave a review - let other readers know what you think

Please share your thoughts on this book with others by leaving a review on the site that you bought it from. If you purchased the book from Amazon, please leave us an honest review on this book's Amazon page. This is vital so that other potential readers can see and use your unbiased opinion to make purchasing decisions, we can understand what our customers think about our products, and our authors can see your feedback on the title that they have worked with Packt to create. It will only take a few minutes of your time, but is valuable to other potential customers, our authors, and Packt. Thank you!

Index

CPSIA information can be obtained
at www.ICGtesting.com
Printed in the USA
FSHW011237081119
63875FS